Test Scoring and Analysis Using SAS®

Ron Cody and Jeffrey K. Smith

support.sas.com/bookstore

The correct bibliographic citation for this manual is as follows: Cody, Ron, and Jeffrey K. Smith. 2014. *Test Scoring and Analysis Using SAS®*. Cary, NC: SAS Institute Inc.

Test Scoring and Analysis Using SAS®

ISBN 978-1-61290-924-0 (Hardcopy)
ISBN 978-1-62959-495-8 (EPUB)
ISBN 978-1-62959-496-5 (MOBI)
ISBN 978-1-62959-497-2 (PDF)

SAS Institute Inc., SAS Campus Drive, Cary, North Carolina 27513-2414.

December 2014

Contents

List of Programs

Chapter 2

Chapter 3

Chapter 4

Chapter 5

Chapter 6

Chapter 7

Chapter 8

Chapter 10

Chapter 11

About This Book

Purpose

This book provides a practical discussion of test scoring, item analysis, test reliability, and other related topics that would be useful to almost any instructor in high school or college who not only wants to score tests and produce rosters, but also wants to investigate the quality of each item and the reliability of the test as a whole.

There are also many agencies that create accreditation tests that also require scoring and item analysis. All of the SAS programs in the book include detailed explanations of how they work. However, even readers not familiar with SAS programming can easily use the scoring and analysis programs.

Prerequisites

The audience for this book is not limited to SAS users—it is anyone who wants to use SAS to score tests, produce class rosters, analyze test items, and measure overall test reliability. But, for people who are familiar with SAS, and this is a large market, many of them have the need to develop test analysis programs, and this book simply lays all of this out for them.

Scope of This Book

Although there are many excellent books on psychometrics on the market, this book is unique in several ways: First, it is basic enough for people without any prior knowledge of test scoring and analysis or who do not have an extensive mathematics background. Secondly, included in this book are complete, pre-packaged programs (called macros in SAS lingo) that can be used by non-programmers. These programs can all be downloaded for free from the author page for this book.

PROC IRT (item response theory) is new to SAS and this is the first book that describes item response theory and how to run PROC IRT.

About the Examples

SAS 9.4 was used to create the programming examples for this book.

Example Code and Data

You can access the example code and data for this book by linking to its author page at http://support.sas.com/publishing/authors. Select the name of the author. Then, look for the cover of this book, and select Example Code and Data to display the SAS programs that are included in this book.

For an alphabetical listing of all books for which example code and data is available, see http://support.sas.com/bookcode. Select a title to display the book's example code.

If you are unable to access the code through the Web site, send e-mail to saspress@sas.com.

Additional Help

Although this book illustrates many analyses regularly performed in businesses across industries, questions specific to your aims and issues may arise. To fully support you, SAS Institute and SAS Press offer you the following help resources:

- About topics covered in this book, contact the author through SAS Press:
  - Send questions by e-mail to saspress@sas.com; include the book title in your correspondence.
  - Submit feedback on the author's page at http://support.sas.com/author_feedback.
- About topics in or beyond this book, post questions to the relevant SAS Support Communities at https://communities.sas.com/welcome.
- SAS Institute maintains a comprehensive website with up-to-date information. One page that is particularly useful to both the novice and the seasoned SAS user is its Knowledge Base. Search for relevant notes in the "Samples and SAS Notes" section of the Knowledge Base at http://support.sas.com/resources.
- Registered SAS users or their organizations can access SAS Customer Support at http://support.sas.com. Here you can pose specific questions to SAS Customer Support: Under *Support*, click *Submit a Problem*. You will need to provide an e-mail address to which replies can be sent, identify your organization, and provide a customer site number or license information. This information can be found in your SAS logs.

Keep in Touch

We look forward to hearing from you. We invite questions, comments, and concerns. If you want to contact us about a specific book, please include the book title in your correspondence.

To Contact the Authors through SAS Press

By e-mail: saspress@sas.com

Via the Web: http://support.sas.com/author_feedback

SAS Books

For a complete list of books available through SAS, visit http://support.sas.com/store/books.

Phone: 1- 800-727-0025

E-mail: sasbook@sas.com

SAS Book Report

Receive up-to-date information about all new SAS publications via e-mail by subscribing to the SAS Book Report monthly eNewsletter. Visit http://support.sas.com/sbr.

Publish with SAS

SAS is recruiting authors! Are you interested in writing a book? Visit http://support.sas.com/saspress for more information.

About These Authors

Ron Cody, EdD, a retired professor from the Robert Wood Johnson Medical School now works as a private consultant and a national instructor for SAS Institute Inc. A SAS user since 1977, Ron's extensive knowledge and innovative style have made him a popular presenter at local, regional, and national SAS conferences. He has authored or co-authored numerous books, such as *Learning SAS by Example: A Programmer's Guide; SAS Statistics by Example, Applied Statistics and the SAS Programming Language, Fifth Edition; The SAS Workbook; The SAS Workbook Solutions; Cody's Data Cleaning Techniques Using SAS, Second Edition; Longitudinal Data and SAS: A Programmer's Guide; SAS Functions by Example, Second Edition*, and *Cody's Collection of Popular Programming Tasks and How to Tackle Them*, as well as countless articles in medical and scientific journals.

Jeffrey Smith, is professor and Associate Dean (Research) in the College of Education at the University of Otago in New Zealand. For 29 years he was on the faculty of Rutgers University, serving the Department of Educational Psychology as professor and chair. From 1988 to 2005 he also served as Head of the Office of Research and Evaluation at the Metropolitan Museum of Art. He has written or edited eight books on educational assessment and statistics, the psychology of aesthetics, and educational psychology. He has published more than 70 research articles and reviews in the field of education, also founding and co-editing a journal, *Psychology of Aesthetics, Creativity, and the Arts*. Smith received his undergraduate degree from Princeton University and his Ph.D. from the University of Chicago.

Learn more about these authors by visiting their author pages, where you can download free book excerpts, access example code and data, read the latest reviews, get updates, and more:
http://support.sas.com/publishing/authors/cody.html
http://support.sas.com/publishing/authors/smith_jeff.html

Acknowledgments

OK, we know that most of you will skip right over this page—but we would like to take a moment to give credit to so many people who were involved in making this book a reality.

The first stage in getting started on our book was to refine our ideas with the SAS Press Editor-and-Chief, Julie Platt, and the acquisitions editor, Shelly Sessoms. It was this back-and-forth dialog that helped refine our ideas and resulted in a book that we hope you will find useful.

Next, we were lucky enough to have two editors: John West started the ball rolling and Sian Roberts made the final goal. Thanks to both of you for your patience and understanding.

One of the most important tasks in the development of a technical book of this nature is to have a talented staff of technical reviewers. This technical review process takes a lot of time and effort and we are grateful to Xinming An, Ben Davidson, Penny Downy, Jill Tao, Cat Truxillo, and Yiu-Fai Yung.

We greatly appreciate the advice and editorial contributions to the chapter on IRT theory from our friend and colleague, Ron Mead. Specifically, he brought objectivity and accuracy to the chapter; any remaining errors will have to be credited to the second author of this text.

Because books now need to be readable on many different e-devices, the Word templates and the associated software needed to produce a book have become quite complicated. Our thanks to Stacey Hamilton who actually understands this stuff and was able to guide us.

Kathy Restivo was our copy editor and she is amazing! It seems that nothing got past her keen eye.

We wish thank Denise Jones and Monica McClain who handled the final formatting edit and made sure that the book will look good in print as well as on the various e-book formats.

Very early on we had the idea for a cover design that incorporated a mark sense answer sheet. Robert Harris, the cover design artist surpassed our expectations and came up with a fantastic cover. Thank you Robert.

Ron Cody and Jeff Smith

Chapter 1: What This Book Is About

Introduction

This book has dual purposes: One is to describe basic (and some advanced) ideas about how tests are evaluated. These ideas start from such simple tasks as scoring a test and producing frequencies on each of the multiple-choice items to such advanced tasks as measuring how well test items are performing (item analysis) and estimating test reliability. You will even find some programs to help you determine if someone cheated on your test. In addition to discussing test theory and SAS programs to analyze tests, we have included a chapter on how to write good test items.

Tests are used in schools to assess competence in various subjects and in professional settings to determine if a person should be accredited (or reaccredited) to some profession, such as becoming a nurse, an EMT, or a physician. Too many of these tests are never evaluated. Why not? Because up until now, easy-to-use programs were not readily available.

A second purpose is to provide you with a collection of programs to perform most of the tasks described in this book. Even without a complete knowledge of SAS programming, you will learn enough to use the included programs to create better tests and assessment instruments. For those who know how to write SAS programs (on a beginning or intermediate level), you will find detailed explanations of how these programs work. Feel free to skip these sections if you like. The last chapter of this book contains listings of programs that you can use to score tests, print student rosters, perform item analysis, and conduct all the tasks that were developed in this book. Along with these listings, you will also find instructions telling you how to run each of the programs.

An Overview of Item Analysis and Test Reliability

Testing is used for an incredibly wide range of purposes in society today. Tests are used for certification into professions, admission into universities and graduate programs, grading at all levels of schooling, formative assessment to help students learn, and classification to determine whether students need special forms of assistance. In each of these settings, it is critical that the use and interpretation of the test scores be valid. Simply put, the test should be doing what it is supposed to do.

Part of what "doing what it is supposed to do" is related to notions of validity and reliability. These concepts are analogous to the concepts of "antique" and "old." In order for something to be an antique, it *has* to be old. But just because something is old doesn't mean it is an antique. Reliability is a necessary but not sufficient condition for being valid. The use and interpretation of a test is valid if it leads to the proper decisions: the right people being certified, the best qualified candidates being selected, grades being fair and accurate, instructional assistance being on target and useful.

What we are going to discuss in this book primarily has to do with the development of tests that are reliable and the analysis of test items to ensure that reliability. We will also discuss the idea of test validity, but test validation research in general is beyond the scope of what we will cover here. We recommend one of these references to get a good idea of what test validity and conducting validation studies is about:

American Educational Research Association, American Psychological Association, & National Council on Measurement in Education (1999). *Standards for educational and psychological testing.* Washington, DC: American Educational Research Association.

Shepard, L. A. (1993). Evaluating test validity. *Review of Research in Education, 19*,405-450.

Kane, M. T. 2006. "Validation." *Educational Measurement, 4th ed*, ed. R. L. Brennan, 17-64. ACE/Praeger Series on Higher Education. MD: Rowman and Littlefield.

The basic idea of reliability has to do with consistency of information. If you tested again, would you get fundamentally the same results? It's like stepping on the scale a second time to make sure that you are really the weight that the scale reported the first time! In order to get that reliability, you need to ensure that all of the items that comprise the test are measuring the same underlying trait, or construct. In this book, you will see how to write items that will make good tests and how to statistically analyze your items using SAS to guide the revision of items and the generation of tests that will have strong reliability.

A Brief Introduction to SAS

This section is intended for those readers who are not familiar with SAS. SAS is many things— a programming language, a collection of statistical procedures, a set of programs to produce a variety of graphs and charts, and an advanced set of tools to provide businesses with advanced business analytics. The programs developed in this book mostly use Base SAS (the programming part of SAS), some graphics programs to provide charts and graphs, and a new SAS procedure to analyze tests using item response theory (IRT).

SAS programs are composed of statements that do the following:

1. Instruct the computer to read data (from a variety of sources such as text files or Excel workbooks)
2. Perform calculations
3. Make logical decisions
4. Output data to files or printers

There are many books that can teach you how to program using SAS. We recommend the following (all available from SAS Institute at www.support.sas.com/publishing):

Delwiche, Laura and Susan Slaughter. 2012. *The Little SAS Book: A Primer, Fifth Edition,* Cary, NC: SAS Press.

Cody, Ron. 2007. *Learning SAS by Example: A Programmer's Guide,* Cary, NC: SAS Press.

SAS Institute Inc. 2013. *SAS 9.4 Language Reference: Concepts*, Cary, NC: SAS Press

SAS Institute Inc. 2013. *SAS 9.4 Macro Language Reference*, Cary, NC: SAS Press.

Chapter 2: Reading Test Data and Scoring a Test

Introduction

This chapter shows you how to read test data from a variety of sources (e.g., text files, CSV files, Excel workbooks) and how to compare the student responses and the answer key to score a test. In addition, you will see how to generate student rosters with or without the inclusion of student names. Finally, you will see how to generate statistics and histograms based on the resulting test scores.

Two popular sources of test data are output from optical mark-sense readers (also called *scanners*) or from data captured online. The answer key information may sometimes be included in the data file or it may be in a separate location. When a scanner is used, it is a common practice to scan the answer key in first, followed by the student responses. You will see examples of how to read data in different formats and how to process the answer key data, depending on how it is stored.

Reading Data from a Text File and Scoring a Test

The programs presented in this section assume that the answer key is stored in the same file as the student responses. Let's get started:

The "test" you are going to score is a 10-question multiple-choice test with possible answer choices of a, b, c, d, and e. This first sample program assumes the answer key and student responses are stored in a text file (sample_test.txt) located in a folder called c:\books\test scoring. Here is a listing of this file:

Listing of File c:\books\test scoring\sample_test.txt

```
000000000 abcedaabed
123456789 abcdeaabed
111111111 abcedaabed
222222222 cddabedeed
333333333 bbdddaaccd
444444444 abcedaabbb
555555555 eecedabbca
666666666 aecedaabed
777777777 dbcaaaabed
888888888 bbccdeebed
999999999 cdcdadabed
```

Notice that the answer key is the first line of the file and that the ID field contains all zeros (you can use any value in the ID field of the answer key or leave it blank). The program presented here ignores the ID field for the answer key. Here is a SAS program that will read data from this text file and produce a listing of student IDs, the raw score (number of items answered correctly), and a percentage score:

Program 2.1: Scoring a Test (Data from a Text File)

```
data score;
   infile 'c:\books\test scoring\sample_test.txt' pad;
   array Ans[10] $ 1 Ans1-Ans10;    ***student answers;
   array Key[10] $ 1 Key1-Key10;    ***answer key;
   array Score[10] Score1-Score10; ***score array 1=right,0=wrong;
   retain Key1-Key10;
   if _n_ = 1 then input @11 (Key1-Key10)($1.);
   input @1  ID $9.
         @11 (Ans1-Ans10)($1.);
   do Item=1 to 10;
      Score[Item] = Key[Item] eq Ans[Item];
   end;
   Raw=sum(of Score1-Score10);
   Percent=100*Raw / 10;
   keep Ans1-Ans10 ID Raw Percent;
   label ID      = 'Student ID'
         Raw     = 'Raw Score'
         Percent = 'Percent Score';
run;

proc sort data=score;
   by ID;
run;

title "Student Scores";
proc print data=score label;
   id ID;
   var Raw Percent;
run;
```

We will explain the salient points of the program in a moment, but first, here is a listing of the output:

Output from Program 2.1

Student Scores

Student ID	Raw Score	Percent Score
111111111	10	100
123456789	8	80
222222222	2	20
333333333	5	50
444444444	8	80
555555555	5	50
666666666	9	90
777777777	7	70
888888888	6	60
999999999	5	50

Each student ID is displayed with the raw test score (the number of items answered correctly) and the score displayed as a percentage.

Explanation of Program 2.1

For those SAS programmers who would like to understand the ins-and-outs of the program, read on. Note: This program uses several statements that are a bit beyond the beginning programmer level.

The program is creating a SAS data set called SCORE.

```
data score;
```

The INFILE statement tells SAS where to find the raw data file. The PAD option is useful in case there are some short records in the file because this option pads each record with blanks

```
infile 'c:\books\test scoring\sample_test.txt' pad;
```

The next three statements create three arrays. Array ANS represents the 10 variables (Ans1-Ans10) that store the student answers. Array KEY represents the 10 answer key variables. Finally, array SCORE represents the 10 score variables: 1=Correct, 0=Incorrect.

```
array Ans[10] $ 1 Ans1-Ans10;    ***student answers;
array Key[10] $ 1 Key1-Key10;    ***answer key;
array Score[10] Score1-Score10; ***score array
```

A RETAIN statement is needed since the answer key is read only once and, without this statement, the 10 key variables would be set back to missing at the top of the DATA step.

```
retain Key1-Key10;
```

The first record in the text file contains the answer key—the remaining records contain the student answers. The SAS created variable _N_ counts iterations of the DATA step. When it is equal to 1, you want to read in the answer key values. These values are stored in the 10 variables Key1 to Key10, starting in column 11.

```
if _n_ = 1 then input @11 (Key1-Key10)($1.);
```

You now read the student IDs starting in column 1 (in this example, the student IDs are 9 characters long) and the 10 student answers (stored in variables Ans1 to Ans10), starting in column 11.

```
input @1  ID $9.
      @11 (Ans1-Ans10)($1.);
```

You now want to compare each of the student answers to the answer key. You do this by writing a DO loop. This loop will iterate 10 times, with the variable Item taking on values from 1 to 10. The first time the loop iterates, Item is equal to 1. The logical expression `Key[1] eq Ans[1]` is true when Ans1 is equal to Key1 and false when Ans1 is not equal to Key1. SAS uses the value 1 for true expressions and 0 for false expressions. Therefore, the value of Score1 will be a 1 when the student answered the question correctly and 0 otherwise. Once the DO loop has completed its 10 iterations, there will be 10 Score variables with values of 1 or 0.

```
do Item=1 to 10;
   Score[Item] = Key[Item] eq Ans[Item];
end;
```

The raw score is the sum of the 10 Score values.

```
Raw=sum(of Score1-Score10);
```

The percentage score is the raw score divided by the number of items on the test (10).

```
Percent=100*Raw / 10;
```

You want to keep the answer key values (to be used in programs that follow), the student ID, the raw score, and the percentage score.

```
keep Ans1-Ans10 ID Raw Percent;
```

A LABEL statement provides labels for selected variables. These labels will be printed in the output.

```
label ID      = 'Student ID'
      Raw     = 'Raw Score'
      Percent = 'Percent Score';
```

The RUN statement ends the DATA step.

```
run;
```

The SORT procedure sorts the data set in increasing order of the student ID.

```
proc sort data=score;
   by ID;
run;
```

Finally, PROC PRINT is used to list the contents of the SCORE data set. The LABEL option tells SAS to use the variable labels and not the variable names in the listing.

```
title "Student Scores".
proc print data=score label;
   id ID;
   var Raw Percent;
run;
```

Reading Space-Delimited Data

The next several sections describe how to read test data from a variety of different formats. You will see how to read space- or comma-delimited data and how to read data from an Excel workbook. It goes without saying that you may wish to skip those sections that demonstrate how to read data from sources you will not be using. On the other hand, if you are a SAS programmer, you may find these sections interesting.

Suppose your data file looks like this:

File c:\books\test scoring\spacedlim.txt

```
a b c e d a a b e d
123456789 a b c d e a a b e d
111111111 a b c e d a a b e d
222222222 c d d a b e d e e d
333333333 b b d d d a a c c d
444444444 a b c e d a a b b b
555555555 e e c e d a b b c a
666666666 a e c e d a a b e d
777777777 d b c a a a a b e d
888888888 b b c c d e e b e d
999999999 c d c d a d a b e d
```

Here, the first row of data contains the answer key and the remaining rows contain an ID and the student answers. Each of the key values and the student answers are separated by blanks. SAS has the ability to read delimited data (it calls the method *list input*) and it is quite straightforward. Because blanks are the default delimiter for SAS list input, you simply list the variable names for each of the delimited data values. The next program reads this data file and the resulting data set is identical to the one created by Program 2.1.

Program 2.2: Reading Space-Delimited Data

```
*Reading Space-Delimited Data;
data read_dlim;
   length ID $ 9;
   infile 'c:\books\test scoring\spacedlim.txt' missover;
   retain Key1-Key10;
   array Key[10] $ 1 Key1-Key10;
   array Ans[10] $ 1 Ans1-Ans10;
   array Score[10] Score1-Score10;
   retain Key1-Key10;
   if _n_ = 1 then input Key1-Key10;
   input ID Ans1-Ans10;
   do Item=1 to 10;
      Score[Item] = Key[Item] eq Ans[Item];
   end;
   Raw=sum(of Score1-Score10);
   Percent=100*Raw / 10;
   keep Ans1-Ans10 ID Raw Percent;
   label ID      = 'Student ID'
         Raw     = 'Raw Score'
         Percent = 'Percent Score';
run;
```

To read space-delimited data, you provide a list of the variables on the INPUT statement in the same order as the variables in the file. When using this method of input, you need to specify the length of character variables since, by default, SAS assigns a length of eight bytes for these variables. In this program, the length of the answer and key variables is defined in the ARRAY statements. To specify the length of the ID variable, you use a LENGTH statement.

The INFILE statement tells the program where to find the input data file. The MISSOVER option is useful when you have a line of data that does not contain as many values as you listed on the INPUT statement. With this option in effect, if there are too few data values in the input data, a missing value is assigned to all the remaining variables. In this example, every line of data had an ID and exactly 10 answers. The MISSOVER option was included just in case there was a line of data with fewer than 10 answers.

The remainder of the program follows the logic of Program 2.1.

Reading Comma-Delimited Data (CSV File)

It is probably even more common to have data values separated by commas. These files can be created using Excel. You only need to make a small change to Program 2.2 to read CSV files.

If you add the option DSD to the INFILE statement, SAS will understand that commas are to be considered as data delimiters instead of spaces. Here is an example of a CSV file that has the answer key as the first line and the student IDs and student answers in the remaining lines:

File c:\books\test scoring\Exceldata.csv

```
a,b,c,e,d,a,a,b,e,d,
123456789,a,b,c,d,e,a,a,b,e,d
111111111,a,b,c,e,d,a,a,b,e,d
222222222,c,d,d,a,b,e,d,e,e,d
333333333,b,b,d,d,d,a,a,c,c,d
444444444,a,b,c,e,d,a,a,b,b,b
555555555,e,e,c,e,d,a,b,b,c,a
666666666,a,e,c,e,d,a,a,b,e,d
777777777,d,b,c,a,a,a,a,b,e,d
888888888,b,b,c,c,d,e,e,b,e,d
999999999,c,d,c,d,a,d,a,b,e,d
```

Here is the SAS program that reads these data:

Program 2.3: Reading Data from a CSV File

```
data readCSV;
   length ID $ 9;
   infile 'c:\books\test scoring\Exceldata.csv' missover dsd;
   retain Key1-Key10;
   array Key[10] $ 1 Key1-Key10;
   array Ans[10] $ 1 Ans1-Ans10;
   array Score[10] Score1-Score10;
   retain Key1-Key10;
   if _n_ = 1 then input Key1-Key10;
   input ID Ans1-Ans10;
   do Item=1 to 10;
      Score[Item] = Key[Item] eq Ans[Item];
   end;
   Raw=sum(of Score1-Score10);
   Percent=100*Raw / 10;
   keep Ans1-Ans10 ID Raw Percent;
   label ID      = 'Student ID'
         Raw     = 'Raw Score'
         Percent = 'Percent Score';
run;
```

Reading Data Directly from an Excel Workbook

This section will show you how to read data directly from an Excel workbook, without first converting it into a CSV file. The method used has a somewhat fancy name—you are going to use a LIBNAME engine. By appropriately identifying the data source, SAS will read the data almost as if it was contained in a SAS data set rather than in an Excel workbook. To demonstrate how this works, your Excel workbook is shown below:

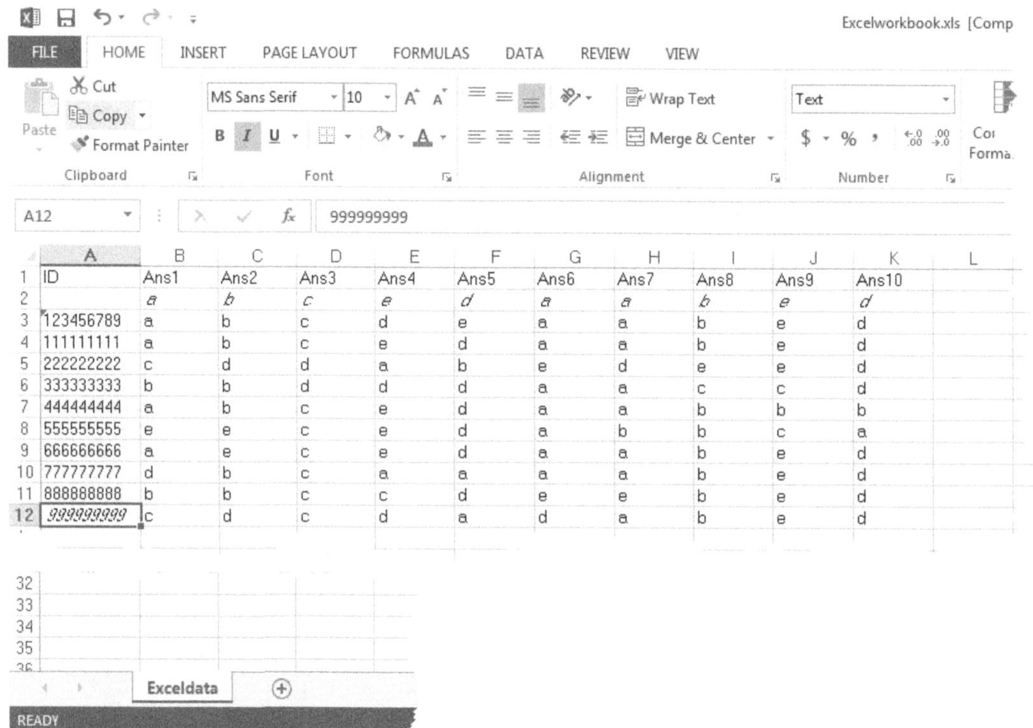

	A	B	C	D	E	F	G	H	I	J	K
1	ID	Ans1	Ans2	Ans3	Ans4	Ans5	Ans6	Ans7	Ans8	Ans9	Ans10
2		a	b	c	e	d	a	a	b	e	d
3	123456789	a	b	c	d	e	a	a	b	e	d
4	111111111	a	b	c	e	d	a	a	b	e	d
5	222222222	c	d	d	a	b	e	d	e	e	d
6	333333333	b	b	d	d	d	a	a	c	c	d
7	444444444	a	b	c	e	d	a	a	b	b	b
8	555555555	e	e	c	e	d	a	b	b	c	a
9	666666666	a	e	c	e	d	a	a	b	e	d
10	777777777	d	b	c	a	a	a	a	b	e	d
11	888888888	b	b	c	c	d	e	e	b	e	d
12	999999999	c	d	c	d	a	d	a	b	e	d

In this workbook, the first row contains the labels ID and Ans1 to Ans10. The next row, starting in the second column (so it lines up with the student answers), is the answer key. Finally, the remaining rows contain the student IDs and the student answers.

Note that the worksheet was named Exceldata, as can be seen in the section at the bottom of the worksheet.

Here is the program to read this worksheet and convert it into the appropriately structured SAS data set:

Program 2.4: Reading Data Directly from an Excel Workbook

```
*Read Excel Workbook;
libname readxl 'c:\books\test scoring\excelworkbook.xls';
data read;
   set readxl.'Exceldata$'n;
   retain Key1-Key10;
   array Key[10] $ 1;
   array Ans[10] $ 1;
   array Score[10] Score1-Score10;
   if _n_ = 1 then do Item = 1 to 10;
      Key[Item] = Ans[Item];
   end;
   drop Item;
   do Item=1 to 10;
      Score[Item] = Key[Item] eq Ans[Item];
   end;
   Raw=sum(of Score1-Score10);
   Percent=100*Raw / 10;
   keep Ans1-Ans10 ID Raw Percent;
   label ID      = 'Student ID'
         Raw     = 'Raw Score'
         Percent = 'Percent Score';
run;
```

You create a library reference (called a *libref* in SAS terminology) using a LIBNAME statement. LIBNAME statements are routinely used to create permanent SAS data sets. When you use them in that context, you simply name a PC or UNIX folder where you want to create your permanent SAS data set. In this program, you use the LIBNAME statement to point to the Excel worksheet you want to read.

The SET statement does two things: First, it names the libref so SAS knows which workbook to read from. Second, the part following the period is the name of the specific worksheet you want to read. You might expect that this should be "Exceldata$." However, that is not a valid SAS name (SAS names must begin with a letter or underscore, followed by letters, digits, or underscores). These headings automatically become variable names in the SAS data set. Space characters or special characters such as the '$' in the worksheet name are not allowed. In order to include the '$' in the worksheet name, you use a *named literal*. That is, you place the name in single (or double) quotes followed by an upper- or lowercase 'n'.

The headings of the worksheet are ID and Ans1-Ans10. However, the first line of data in the worksheet contains the answer key, not student answers. Therefore, you need to set the values in the first line of the worksheet to the 10 Key variables. When you are reading the first line of the file, the internal variable _N_ is equal to 1. In that case, you set the 10 Key variables equal to the 10 Ans variables using a DO loop. From that point on, the program logic is the same as described in Program 2.1.

Reading an Answer Key from a Separate File

If you have the answer key stored in a separate file, you need to make a small change to the scoring program in order to score your test. As an example, suppose your answer key is stored as a CSV text file called c:\books\test scoring\key.txt and the student data is stored in the same format as Program 2.1 and called answer.txt. To be sure this is clear, here are the listings of these two files:

File c:\books\test scoring\key.txt

```
a,b,c,e,d,a,a,b,e,d
```

File c:\books\test scoring\answer.txt

```
123456789 abcdeaabed
111111111 abcedaabed
222222222 cddabedeed
333333333 bbdddaaccd
444444444 abcedaabbb
555555555 eecedabbca
666666666 aecedaabed
777777777 dbcaaaabed
888888888 bbccdeebed
999999999 cdcdadabed
```

Here is one version of the scoring program that will read data from these two files and compute student scores (a brief explanation will follow):

Program 2.5: Program to Read the Answer Key from a Separate File

```
*Program to read the answer key from a separate file;
data score2;
   if _n_ = 1 then do;
      infile 'c:\books\test scoring\key.txt' dsd;
      input (Key1-Key10)(: $1.);
   end;
   infile 'c:\books\test scoring\answer.txt' pad;
   array ans[10] $ 1 Ans1-Ans10;    ***student answers;
   array key[10] $ 1 Key1-Key10;    ***answer key;
   array score[10] Score1-Score10; ***score array 1=right,0=wrong;
   retain Key1-Key10;
   input @1  ID $9.
         @11 (Ans1-Ans10)($1.);
   do Item=1 to 10;
      Score[Item] = Key[Item] eq Ans[Item];
   end;
   Raw=sum(of Score1-Score10);
   Percent=100*Raw / 10;
   keep Ans1-Ans10 ID Raw Percent;
   label ID      = 'Student ID'
         Raw     = 'Raw Score'
         Percent = 'Percent Score';
run;
```

On the first iteration of the DATA step, you read the 10 answer key values from the key.txt file. Once you have read the answer key, you use another INFILE statement to switch the input to the student answer file. The remainder of the program is the same as Program 2.1.

Modifying the Program to Score a Test of an Arbitrary Number of Items

The next step towards making the scoring program more useful is to modify it to score a test containing any number of items. To accomplish this goal, you replace the constant value of 10 in Program 2.1with a macro variable. For those readers who are not familiar with SAS macros, the concept is analogous to the find-and-replace facility in most word processors. In the case of SAS macros, macro variables (variable names that start with an ampersand) are replaced with values, usually supplied in a call to the macro. Question: How do you call a macro? Answer: Shout MACRO! You will see how to call (execute) the macro following the listing.

The following macro scores a test consisting of an arbitrary number of items. To make the macro even more flexible, you can enter the length of the ID as well as the starting position for the answer key and the student answers. Here it is:

Program 2.6: Presenting a Macro to Score a Test of Any Length

```
%macro score(File=,      /*Name of the file containing the answer
                           key and the student answers       */
             Length_id=, /*Number of bytes in the ID         */
             Start=,     /*Starting column of student answers */
             Nitems=     /*Number of items on the test       */);

   data score;
      infile "&File" pad;
      array Ans[&Nitems] $ 1 Ans1-Ans&Nitems;      ***student answers;
      array Key[&Nitems] $ 1 Key1-Key&Nitems;      ***Answer Key;
      array Score[&Nitems] Score1-Score&Nitems;    ***score array
                                                   1=right,0=wrong;
      retain Key1-Key&Nitems;
      if _n_= 1 then input @&Start (Key1-Key&Nitems)($1.);
      input @1  ID $&Length_ID..
            @&start (Ans1-Ans&Nitems)($1.);
      do Item = 1 to &Nitems;
         Score[Item] = Key[Item] eq Ans[Item];
      end;
      Raw=sum (of Score1-Score&Nitems);
      Percent=100*Raw / &Nitems;
      keep Ans1-Ans&Nitems ID Raw Percent;
      label ID      = 'Student ID'
            Raw     = 'Raw score'
            Percent = 'Percent score';
   run;

   proc sort data=score;
      by ID;
   run;

%mend score;
```

You don't need to be a SAS programmer or understand how to write SAS macros to use this macro to score tests. All you need to know is how to specify the variables required in the call. To call a SAS macro, you enter a percent sign followed by the macro name, followed by the values needed for the macro variables.

As an example, suppose you want to score a test where the raw data file is called sample_test.txt, stored in a folder called c:\books\test scoring. You have nine-digit ID numbers, the student answers start in column 11, and the test contains 10 items. The macro call looks like this:

```
*Calling the macro;
%score(File=c:\books\test scoring\sample_test.txt,
       Start=11,
       Length_ID=9,
       Nitems=10)
```

Notice that this macro call does not end in a semi-colon. When you run this macro, you will obtain the same output as with the non-macro versions described earlier. The advantage of converting the program into a SAS macro is that you don't have to rewrite the program to change the input file name, the length of the ID, the location of the student answers, or the number of items on the test each time you want to score a test.

If you want to see a listing of student IDs, raw scores, and percentage scores, you can use PROC PRINT, like this:

Program 2.7: Using PROC PRINT to List Student IDs and Scores

```
title "Listing of Student IDs and Test Scores";
proc print data=score;
   id ID;
   var Raw Percent;
run;
```

The resulting listing is shown next:

Output from Program 2.7

Student Scores

ID	Raw	Percent
111111111	10	100
123456789	8	80
222222222	2	20
333333333	5	50
444444444	8	80
555555555	5	50
666666666	9	90
777777777	7	70
888888888	6	60
999999999	5	50

The list of IDs is in order because the macro program included a sort on ID.

Displaying a Histogram of Test Scores

One way to display a histogram of test scores is with PROC UNIVARIATE. This procedure can also output some useful statistics on the distribution of scores. The program shown next produces some commonly used measures of central tendency and dispersion as well as a histogram of test scores:

Program 2.8: Program to Produce Statistics and Histograms of Test Scores

```
*Program to Produce Statistics and Histograms of Test Scores;
%score(File=c:\books\test scoring\stat_test.txt,
       Start=11,
       Length_ID=9,
       Nitems=56)

title "Test Statistics and Histograms";
proc univariate data=score;
   id ID;
   var Percent;
   histogram / midpoints=30 to 100 by 5;
run;
```

The %SCORE macro, described earlier in this chapter, was used to score a biostatistics test containing 56 items. (Note that the IDs from the original test were replaced by random digits.) If you include an ID statement with PROC UNIVARIATE, a section of the output that lists the five highest and five lowest values will use the ID variable to identify these observations. You use a VAR statement to specify which variables you want to analyze. Finally, the HISTOGRAM statement produces a histogram. In this example, the range of scores was set from 30 to 100 at intervals of 5 points.

Here are sections of the output, with a short description below each section:

Output from Program 2.8 (Moments)

Test Statistics and Histograms

The UNIVARIATE Procedure
Variable: Percent (Percent score)

Moments			
N	137	Sum Weights	137
Mean	72.9927007	Sum Observations	10000
Std Deviation	10.795919	Variance	116.551867
Skewness	-0.9101907	Kurtosis	0.90686028
Uncorrected SS	745778.061	Corrected SS	15851.0539
Coeff Variation	14.790409	Std Error Mean	0.92235761

This portion of the output shows you the number of students taking the test (N), the mean, and the standard deviation. The skewness and kurtosis values are useful in deciding if the distribution of scores is normally distributed. *Skewness* is a statistic that indicates if the distribution is skewed to the right (positive value) or to the left (negative value). A value of 0 indicates that the distribution is not skewed. A normal distribution would have a skewness value of 0. The *kurtosis statistic* measures if the distribution is too peaked or too flat, relative to a normal distribution. A normal distribution would have a kurtosis value of 0. Most of the other statistics in this table are not of interest in analyzing a test.

Output from Program 2.8 (Basic Statistical Measures)

Basic Statistical Measures			
Location		Variability	
Mean	72.99270	Std Deviation	10.79592
Median	75.00000	Variance	116.55187
Mode	76.78571	Range	58.92857
		Interquartile Range	12.50000

This portion of the output displays most of the common measures of central tendency and dispersion. As you will see in a later chapter on test reliability, larger values for the standard deviation tend to result in tests that are more reliable.

Output from Program 2.8 (Tests for Location)

<table>
<tr><th colspan="5">Tests for Location: Mu0=0</th></tr>
<tr><th>Test</th><th colspan="2">Statistic</th><th colspan="2">p Value</th></tr>
<tr><td>Student's t</td><td>t</td><td>79.13709</td><td>Pr > |t|</td><td><.0001</td></tr>
<tr><td>Sign</td><td>M</td><td>68.5</td><td>Pr >= |M|</td><td><.0001</td></tr>
<tr><td>Signed Rank</td><td>S</td><td>4726.5</td><td>Pr >= |S|</td><td><.0001</td></tr>
</table>

This portion of the output displays statistical tests for the null hypothesis that the mean test score is 0 and, therefore, not useful in this context.

Output from Program 2.8 (Quantiles)

Quantiles (Definition 5)	
Quantile	Estimate
100% Max	92.8571
99%	91.0714
95%	87.5000
90%	85.7143
75% Q3	80.3571
50% Median	75.0000
25% Q1	67.8571
10%	57.1429
5%	50.0000
1%	44.6429
0% Min	33.9286

You can use the table of quantiles to give you more information on the distribution of test scores. In this table, you can see that the median score was 75, with the 25th and 75th percentiles equal to 67.8571 and 80.3571, respectively. It also lets you see the highest and lowest scores (91.0714 and 33.9286) on the test.

Output from Program 2.8 (Extreme Observations)

Extreme Observations					
Lowest			Highest		
Value	ID	Obs	Value	ID	Obs
33.9286	461148393	68	89.2857	205820635	38
44.6429	045841843	6	89.2857	539767715	81
46.4286	102910565	19	91.0714	311170035	46
48.2143	818211122	115	91.0714	383258460	54
48.2143	719581534	105	92.8571	418926812	57

The final tabular output from PROC UNIVARIATE shows you the five highest and five lowest scores on the test, along with the IDs of those students.

Output from Program 2.8 (Histogram)

The UNIVARIATE Procedure

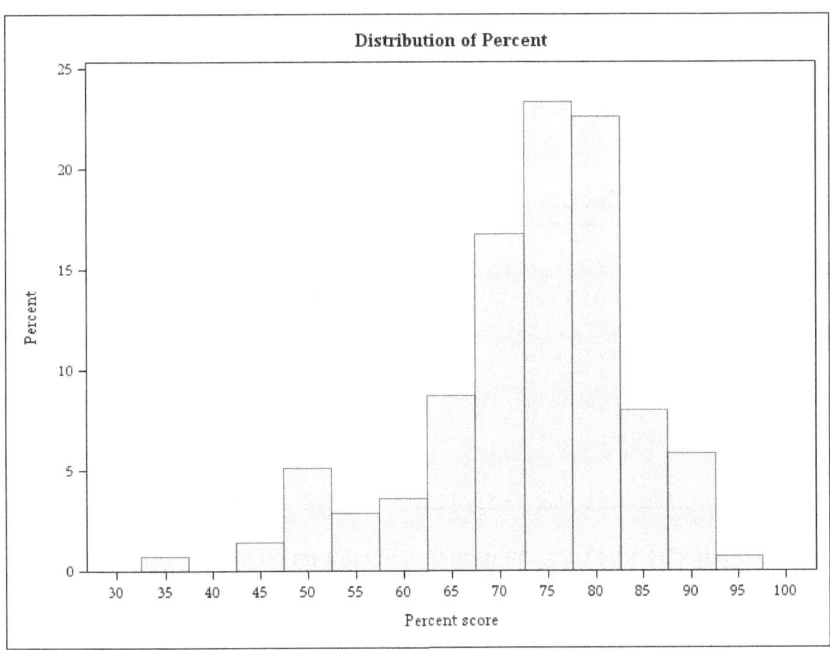

The HISTOGRAM statement produced this rather attractive histogram of test scores. You can change the midpoints easily by choosing different values on the MIDPOINTS option (following the slash on the HISTOGRAM statement).

Matching Student Names with Student IDs

The output from Program 2.1 is useful for posting scores, but you will most likely want a roster showing the student ID, the corresponding student name, and the test scores (and, later, other information). In order to accomplish that goal, you need a file containing student names and IDs. This could be a text file, an Excel worksheet, or some other type of file. For this example, let's assume you have a text file containing the student name, followed by a comma, followed by the student ID. For example, your file might look something like this:

File c:\books\test scoring\student_name.txt

```
Fred Flintstone,11111111
Sandy Lynch,12345678
Julia S. Chien,22222222
Jay Pena,33333333
Raul Heizer,44444444
Janet Friend,55555555
Sandy Lowery,66666666
Carl Weber,77777777
Rube Goldberg,88888888
Casey McDonald,99999999
```

This is a CSV file that you can read using the DSD INFILE option described earlier in this chapter. The program that follows reads the names and IDs from this file and then sorts the file in ID order. The sort is necessary so that you can merge it with the data set containing the student IDs and test scores. Here is the program:

Program 2.9: Creating a SAS Data Set Containing Student Names and IDs

```
data student_name;
   length Name $ 15 ID $ 9 Last_Name $ 10;
   infile 'c:\books\test scoring\student.txt' dsd;
   input Name ID;
      Last_Name = scan(Name,-1," ");
run;

proc sort data=student_name;
   by ID;
run;
```

Since this is a comma-delimited (CSV) file, you use the DSD option on the INFILE statement. Because the Name variable contains both a first and last name, the SCAN function is used to extract the last name (so that it can be used later to provide a roster in alphabetical order by student last name). For those of you who are interested in the programming details, the SCAN function takes a character value and breaks it apart into "words." Words is in quotes since you can specify delimiters other than blanks when using this function. The first argument of the function is the string you want to parse (take apart); the second argument is which word you want. If you use a negative number for this argument, the function scans from right to left. That is particularly useful when you have a first name, possibly a middle initial, followed by a last name. The right-most word is always the last name. The third argument to the SCAN function is a list of delimiters. In this example, you want to treat blanks as word delimiters. One final word about the SCAN function: If you don't define a length for the word you are extracting from the character value, SAS assigns a default length of 200. Therefore, whenever you use this function, you want to use a LENGTH statement to assign a length to the extracted word.

You use PROC SORT to sort the student data set in ID order. The BY statement tells the sort which variable or variables you want to use for the sort.

Here is a listing of the student data set:

Output from Program 2.9

Listing of Data Set STUDENT_NAME

Name	ID	Last_Name
Fred Flintstone	111111111	Flintstone
Sandy Lynch	123456789	Lynch
Julia S. Chien	222222222	Chien
Jay Pena	333333333	Pena
Raul Heizer	444444444	Heizer
Janet Friend	555555555	Friend
Sandy Lowery	666666666	Lowery
Carl Weber	777777777	Weber
Rube Goldberg	888888888	Goldberg
Casey McDonald	999999999	McDonald

You now have a SAS data set containing the student's first name, ID, and last name.

You can merge it with the student roster data stored in the data set SCORE. Here is the program:

Program 2.10: Adding Student Names to the Roster

```
data roster_with_names;
   merge score(keep=ID Raw Percent in=in_score) student_name;
   by ID;
   if in_score;
run;

proc sort data=roster_with_names;
   by Last_Name;
run;

title "Listing of Roster with Names";
proc print data=roster_with_names;
   id Name;
   var ID Raw Percent;
run;
```

The DATA step uses a MERGE statement to add the student names associated with each student ID. Because the STUDENT_NAME data set may contain names and IDs of student who did not take the test, you use the IN= data set option to create a temporary variable called IN_SCORE. This variable has a value of true whenever the merge operation finds an ID in the SCORE data set and of false otherwise. The result is a listing of all the students who took the test, along with their names.

Once the merge is completed, you re-sort the data set so that the final listing is ordered by last name. You now have a listing that looks like this:

Output from Program 2.10

Listing of Roster with Names

Name	ID	Raw	Percent
Julia S. Chien	222222222	2	20
Fred Flintstone	111111111	10	100
Janet Friend	555555555	5	50
Rube Goldberg	888888888	6	60
Raul Heizer	444444444	8	80
Sandy Lowery	666666666	9	90
Sandy Lynch	123456789	8	80
Casey McDonald	999999999	5	50
Jay Pena	333333333	5	50
Carl Weber	777777777	7	70

Creating a Fancier Roster Using PROC REPORT

Although the report displayed above is fine for most purposes, you can use PROC REPORT to customize the appearance. Here is an example:

Program 2.11: Creating a Fancier Roster Using PROC REPORT

```
title "Student Roster in Alphabetical Order";
proc report data=roster_with_names nowd headline;
   columns Name ID Raw Percent;
   define Name / "Student Name" width=15;
   define ID / "Student ID" width=10;
   define Raw / "Raw Score" width=5;
   define Percent / "Percent Score" width=7;
run;
```

This program produces a more customizable report than the one produced using PROC PRINT. You use a COLUMNS statement to list the names of the variables you want to include in your report and a DEFINE statement for each of these variables. Following a slash on the DEFINE statement, you provide labels and widths. The resulting report is shown below:

Output from Program 2.11

Student Roster in Alphabetical Order

Student Name	Student ID	Raw Score	Percent Score
Julia S. Chien	222222222	2	20
Fred Flintstone	111111111	10	100
Janet Friend	555555555	5	50
Rube Goldberg	888888888	6	60
Raul Heizer	444444444	8	80
Sandy Lowery	666666666	9	90
Sandy Lynch	123456789	8	80
Casey McDonald	999999999	5	50
Jay Pena	333333333	5	50
Carl Weber	777777777	7	70

Exporting Your Student Roster to Excel

You may be happy enough to have your student roster in a SAS data set, but you may want to export it to an Excel workbook. You can do this by using the Export Wizard, part of the SAS Display Manager; by using PROC EXPORT; or by using a LIBNAME engine. Here is an example using PROC EXPORT:

Program 2.12: Exporting Your Roster to an Excel Workbook

```
proc export data= work.roster_with_names
            outfile= "c:\books\test scoring\roster.xls"
            dbms=excel replace;
            sheet="class";
run;
```

You use the OUTFILE= option to specify the name and location of the Excel workbook, the DBMS= option to specify that you want to export to Excel, and the REPLACE option to overwrite an existing workbook if it already exists. Finally, the SHEET= option lets you enter a name for the specific worksheet you are creating. Below is a picture of the Excel spreadsheet produced by Program 2.12:

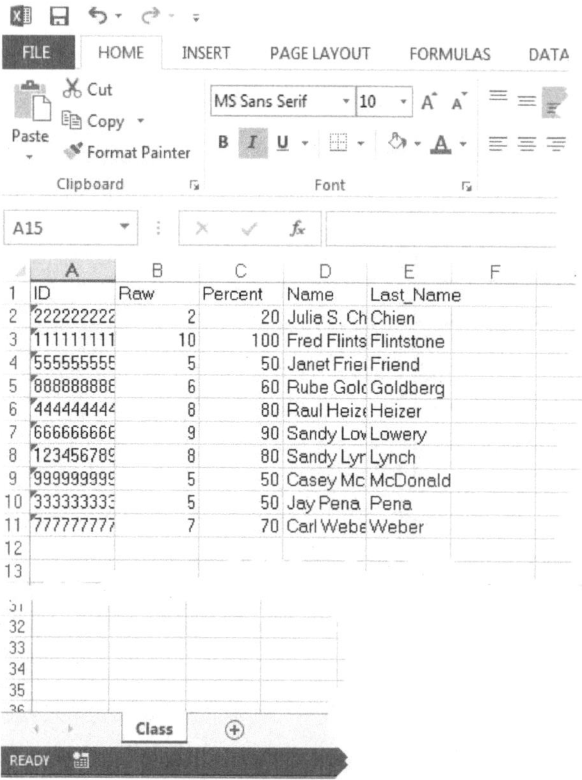

	A	B	C	D	E	F
1	ID	Raw	Percent	Name	Last_Name	
2	222222222	2	20	Julia S. Ch	Chien	
3	111111111	10	100	Fred Flints	Flintstone	
4	555555555	5	50	Janet Frie	Friend	
5	888888888	6	60	Rube Gold	Goldberg	
6	444444444	8	80	Raul Heize	Heizer	
7	666666666	9	90	Sandy Low	Lowery	
8	123456789	8	80	Sandy Lyr	Lynch	
9	999999999	5	50	Casey Mc	McDonald	
10	333333333	5	50	Jay Pena	Pena	
11	777777777	7	70	Carl Webe	Weber	

Conclusion

You have seen how to read test data from a variety of sources and how to write SAS program to score a test. Because it is inconvenient to write a new program every time you need to score a test, you saw how a SAS macro can be used to automate this process. In Chapter 11, you will see a complete list of test scoring and test analysis macros developed in this book.

You also saw how you can use SAS procedures such as UNIVARIATE, PRINT, and REPORT to display test results. Finally, you saw a way to export test data from a SAS data set to an Excel workbook.

Chapter 3: Computing and Displaying Answer Frequencies

Introduction

One of the first steps in analyzing your multiple-choice tests is to display frequencies for each of the item choices. You can do this very simply using PROC FREQ. However, it is useful to see which of the answer choices is the correct answer, along with the frequency information. In this chapter, you will see several ways to accomplish this goal, including a SAS macro that automates the task. Finally, you will see how to display answer frequencies in graphical form.

Displaying Answer Frequencies (in Tabular Form)

To display the answer frequencies for your test, just run PROC FREQ, as demonstrated next:

Program 3.1: Using PROC FREQ to Display Answer Frequencies

```
title "Answer Frequencies";
proc freq data=score;
   tables Ans1-Ans10 / nocum;
run;
```

PROC FREQ can produce one-way (row dimension), two-way (row by column), and three-way (row by column by page) frequency tables. You use a TABLES statement to specify the list of variables for which you want to compute frequencies. In this program, you are asking for frequencies for the 10 answer variables. NOCUM is an option on the TABLES statement—it indicates that you do not want cumulative frequencies or cumulative percentages. All statement options in SAS follow a slash, as demonstrated here. Partial output (frequencies for Ans1 and Ans2) from this program is displayed next:

Output from Program 3.1

Answer Frequencies

The FREQ Procedure

Ans1	Frequency	Percent
a	4	40.00
b	2	20.00
c	2	20.00
d	1	10.00
e	1	10.00

Ans2	Frequency	Percent
b	6	60.00
d	2	20.00
e	2	20.00

While this is useful, it would be helpful to identify the correct answer for each item in the frequency output. That is the subject of the next section.

Modifying the Program to Display the Correct Answer in the Frequency Tables

This section is primarily included for the interested SAS programmer. For those readers who simply want to use these programs to score and analyze tests, the ideas developed in this program are included in the collection of scoring and analysis macros described in Chapter 11.

The key to being able to display the correct answer in the frequency tables is to make the answer variables (Ans1-Ans10 in this example) two bytes long instead of one. To identify the correct answer for each item, you add an asterisk in the second byte of the correct answer for each item. This way, when you request answer frequencies, the correct answer is identified. Here is the program:

Program 3.2: Modifying the Program to Display the Correct Answer to Each Item

```
data score;
   infile 'c:\books\test scoring\sample_test.txt' pad;
   array Ans[10] $ 2 Ans1-Ans10;      ***student answers;
   array Key[10] $ 1 Key1-Key10;      ***Answer Key;
   array Score[10] 3 Score1-Score10; ***score array 1=right,0=wrong;
   retain Key1-Key10;
   if _n_=1 then input @11 (Key1-Key10)($1.);
   input @1  ID $9.
         @11 (Ans1-Ans10)($1.);
   do Item = 1 to 10;
      if Key[Item] eq Ans[Item] then do;
      Score[Item] = 1;
      substr(Ans[Item],2,1)='*';
      ***place an asterisk next to correct answer;
      end;
      else Score[Item] = 0;
   end;
   Raw=sum (of Score1-Score10);
   Percent=100*Raw / 10;
   keep Ans1-Ans10 ID Raw Percent;
   label ID      = 'Student ID'
         Raw     = 'Raw score'
         Percent = 'Percent score';
run;
```

You define the 10 answer variables as two-byte character variables (the ARRAY statement does this). As you score the items, if the student answer matches the answer key, you use the SUBSTR function (on the left-hand side of the equal sign) to place an asterisk in the second position of the variable. Because this use of the SUBSTR function is somewhat esoteric, here is a more detailed explanation:

Normally, the SUBSTR function extracts substrings from an existing string. The first argument of this function is the string of interest, while the second and third arguments are the starting position and the length (optional) of the substring you want to extract, respectively. When used on the left-hand side of the equal sign, this function places characters in an existing string. The definition of the three arguments when the SUBSTR function is used on the left-hand side of the equal sign is identical to its normal usage, except now the string on the right-hand side of the equal sign is placed in the existing string with the appropriate starting position and length.

When you run PROC FREQ, the output displays the correct answer to each item. Below are the first two frequency tables produced by PROC FREQ:

Output from Program 3.2

Answer Frequencies with the Correct Answer Identified

The FREQ Procedure

Ans1	Frequency	Percent
a*	4	40.00
b	2	20.00
c	2	20.00
d	1	10.00
e	1	10.00

Ans2	Frequency	Percent
b*	6	60.00
d	2	20.00
e	2	20.00

You now see the correct answer (indicated by the asterisk) for each item.

Developing an Automated Program to Score a Test and Produce Item Frequencies

The following macro scores a test consisting of an arbitrary number of items as well as computing answer frequencies. Here it is:

Program 3.3: Converting the Test Scoring Program into a Macro

```
%macro ans_freq(Dsn=,    /*Name of the SAS data set you want to create */
                File=,   /*Name of the file containing the answer
                           key and the student answers              */
                Length_id=, /*Number of bytes in the ID              */
                Start=,     /*Starting column of student answers     */
                Nitems=     /*Number of items on the test
*/);
   data &Dsn;
      infile "&File" pad;
      array Ans[&Nitems] $ 2 Ans1-Ans&Nitems;     ***student answers;
      array Key[&Nitems] $ 1 Key1-Key&Nitems;     ***Answer Key;
      array Score[&Nitems] 3 Score1-Score&Nitems; ***score array
                                                  1=right,0=wrong;
      retain Key1-Key&Nitems;
      if _n_= 1 then input @&Start (Key1-Key&Nitems)($1.);
      input @1  ID $&Length_ID..
            @&start (Ans1-Ans&Nitems)($1.);
      do Item = 1 to &Nitems;
         if Key[Item] eq Ans[Item] then do;
         Score[Item] = 1;
         substr(Ans[Item],2,1)='*';
         ***place an asterisk next to correct Answer;
         end;
         else Score[Item] = 0;
      end;
      Raw=sum (of Score1-Score&Nitems);
      Percent=100*Raw / &Nitems;
      keep Ans1-Ans&Nitems ID Raw Percent;
      label ID      = 'Student ID'
            Raw     = 'Raw score'
            Percent = 'Percent score';
   run;

   proc sort data=&Dsn;
      by ID;
   run;

   title "Student Scores";
   proc print data=&Dsn label;
      id ID;
      var Raw Percent;
   run;

   title "Answer frequencies";
   proc freq data=&Dsn;
      tables Ans1-Ans&Nitems / nocum;
   run;
%mend ans_freq;
```

To use this macro, you need to supply the name of the data set you want to create, the name of the text file containing the answer key and the student answers, the length of the ID, the starting column for the answer values, and the number of items on the test. This program assumes that the student ID starts in column 1. If this is not the case, it is a simple matter to replace the input pointer (@1) for the student ID to another value. One other feature of this program is the array Score, which holds the scored values for each item. It also sets the length of these numeric variables to three bytes (the minimum length allowed). This is OK because the value of these variables is either 0 or 1, so using three bytes instead of the default value of eight bytes saves some storage space and speeds up the processing a tiny bit. This is a minor point and you can leave the length of the score variables at eight if you wish.

Suppose you want to score a test where the raw data file is called sample_test.txt, stored in a folder called c:\books\test scoring. You have nine-digit ID numbers, the student answers start in column 11, and the test contains 10 items. The macro call looks like this:

```
*Calling the macro;
%ans_freq(Dsn=item_freq,
          File=c:\books\test scoring\sample_test.txt,
          Start=11,
          Length_ID=9,
          Nitems=10)
```

When you run this macro, you will obtain the same output as with the non-macro versions described earlier.

Displaying Answer Frequencies in Graphical Form

You may wish to see the frequency data in graphical form. This is easily accomplished by using PROC SGPLOT. PROC SGPLOT (SG stands for *statistical graphics*), along with several other SG procedures, became available with version 9.3 (and is included with Base SAS). Prior to this, you needed to license SAS/GRAPH software to produce graphical output.

You can request horizontal or vertical bar charts with HBAR or VBAR statements in PROC SGPLOT. However, because either one of these statements only allows you to plot a single variable, it would be tedious to request bar charts for all the items in a test (even if the test is relatively short). If you restructure the data set, you can make the PROC SGPLOT statements simpler. To help you understand this restructuring process, data set ITEM_FREQ is listed below:

Listing of Data Set ITEM_FREQ

Answer frequencies

ID	Ans1	Ans2	Ans3	Ans4	Ans5	Ans6	Ans7	Ans8	Ans9	Ans10	Raw	Percent
111111111	a*	b*	c*	e*	d*	a*	a*	b*	e*	d*	10	100
123456789	a*	b*	c*	d	e	a*	a*	b*	e*	d*	8	80
222222222	c	d	d	a	b	e	d	e	e*	d*	2	20
333333333	b	b*	d	d	d*	a*	a*	c	c	d*	5	50
444444444	a*	b*	c*	e*	d*	a*	a*	b*	b	b	8	80
555555555	e	e	c*	e*	d*	a*	b	b*	c	a	5	50
666666666	a*	e	c*	e*	d*	a*	a*	b*	e*	d*	9	90
777777777	d	b*	c*	a	a	a*	a*	b*	e*	d*	7	70
888888888	b	b*	c*	c	d*	e	e	b*	e*	d*	6	60
999999999	c	d	c*	d	a	d	a*	b*	e*	d*	5	50

You want to restructure this data set so that each item choice is stored in a separate observation, along with the item number. For example, the first observation in the ITEM_FREQ data set looks like this:

```
   ID      Ans1 Ans2 Ans3 Ans4 Ans5 Ans6 Ans7 Ans8 Ans9 Ans10 Raw Percent

123456789  a*   b*   c*   d    e    a*   a*   b*   e*   d*    8     80
```

You want the first few observations in the restructured data set to look like this:

```
Item    Choice

  1       a*
  2       b*
  3       c*
  4       d*
  5       e*
  6       a*
  7       a*
```

You can compute answer frequencies or create a bar chart of choices for each item if you include Item as a BY variable. The program to restructure the data set where one observation contains all of the items on the test to a data set with one observation per item is shown next:

Program 3.4: Restructuring the ITEM_FREQ Data Set to be Used in PROC SGPLOT

```
data item_choice;
   set item_freq;
   array Ans[10];
   do Item = 1 to 10;
      Choice = Ans[Item];
      output;
   end;
   keep Item Choice;
run;

proc sort data=item_choice;
   by Item;
run;
```

This program reads in one observation from data set ITEM_FREQ and writes out 10 observations in the ITEM_CHOICE data set, with variables Item and Choice. It is important to understand that the OUTPUT statement is inside the DO loop, giving you 10 observations in the ITEM_CHOICE data set for each observation in the ITEM_FREQ data set. If you want to use Item as a BY variable, you need the data set to be sorted by Item. That is the purpose of the final PROC SORT in this program. Here is a listing of the first few observations:

Listing of the First Few Observations in Data Set ITEM_CHOICE

Listing of Data Set ITEM_CHOICE

Item	Choice
1	a*
1	a*
1	c
1	b
1	a*
1	e
1	a*
1	d
1	b
1	c
2	b*
2	b*
2	d
2	b*

You can now request bar charts using PROC SGPLOT as follows:

Program 3.5: Using PROC SGPLOT to Produce Bar Charts for Each Item

```
title "Answer Frequencies";
proc sgplot data=item_choice;
   by Item;
   hbar Choice;
quit;
```

Item is specified on the BY statement and Choice is specified on the HBAR statement. The result is a horizontal bar chart of Choice for each value of Item. The first plot looks like this:

Output from Program 3.5

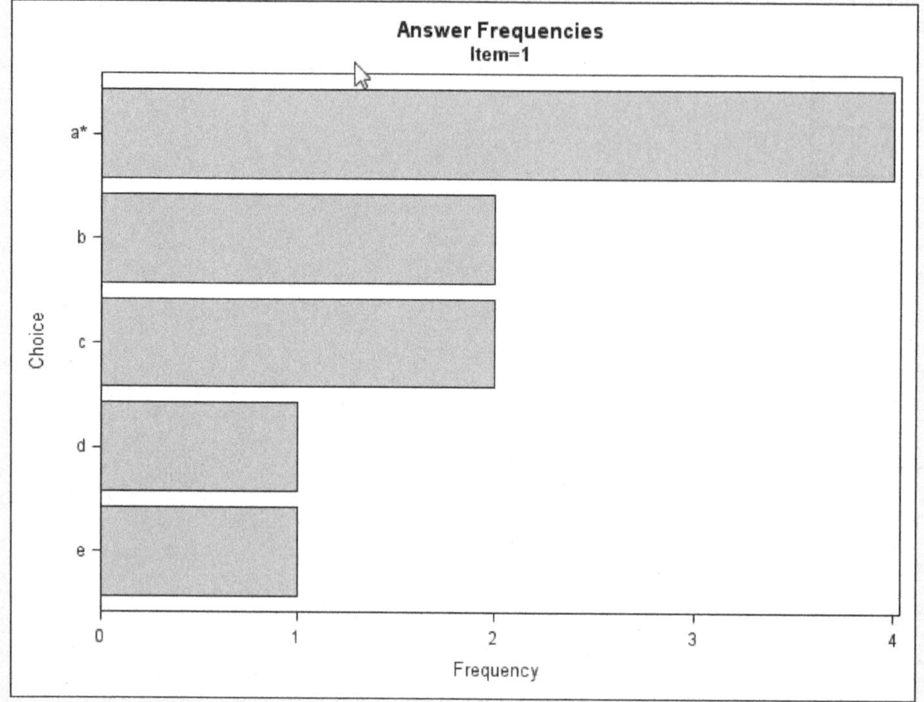

Here you see the answer frequencies for item one in graphical form. In Chapter 5, you will see how to combine answer frequencies with other item statistics, such as item difficulty and item discrimination scores, in a single table.

Conclusion

One of the first steps in evaluating multiple-choice items is to examine the frequency distribution of student responses. With a little innovative programming, you can flag the correct answer in your results. You may want to rewrite or re-evaluate items where too few students answered the item correctly. In addition, you may want to rethink the text of your distractors (the incorrect answer choices) when the frequencies for these choices are too low. Tabular output using PROC FREQ is probably the simplest and most useful way to display answer frequencies; however, you may prefer graphical output, as described in the last section of this chapter.

Chapter 4: Checking Your Test Data for Errors

Introduction

All of the programs developed so far have assumed that there are no errors in the test data. This is an unrealistic assumption. In the real world, it is likely that your student data contains errors. For example, if you are using a mark-sense reader (scanner), the student may have filled in more than one answer choice. More sophisticated scanners include darkest mark discrimination software that attempts to choose the darker mark if more than one answer is selected. This is useful when there are stray marks on the sheet or when the student fails to erase an item completely. Another possible error occurs when a student marks a selection so lightly that it is interpreted as an unanswered item.

Other possible errors include incorrect student IDs (in which case a matching name will not be found in the student name file). Because scanners can jam (and they do), you may wind up with multiple entries for a single student.

Other sources of test data (entered by hand, from the web, etc.) are also subject to data errors. Most of the programs described in this book will either produce strange-looking results or fail completely if certain types of errors in your data remain uncorrected.

You will see ways to identify and fix all of these errors with the SAS programs developed in this chapter.

Detecting Invalid IDs and Answer Choices

Suppose you are using a scanner that does not support darkest mark discrimination. If a student marks more than one choice, you will usually wind up with an ASCII character other than A, B, C, D, or E. One way to detect these errors is with the following short SAS program. (Note: This program, as well as other programs in this book, are available in pre-packaged form in Chapter 11 for use by SAS programmers and other programmers alike.) A file called data_errors.txt was created using the original file sample_test.txt with three errors introduced: one invalid student ID, one answer choice not equal to A through E, and one blank answer. Here is a listing of file data_errors.txt:

Listing of File data_errors.txt

```
000000000 abcedaabed
123456789 a cdeaabed
111111111 abcedaabed
222222222 cddabedeed
333333333 bbdddaaccd
444444444 abce?aabbb
555555555 eecedabbca
666666666 aecedaabed
777x77777 dbcaaaabed
888888888 bbccdeebed
999999999 cdcdadabed
```

You can run the following program to identify the errors in this file:

Program 4.1: Detecting Invalid IDs and Answer Choices

```
title "Checking for Invalid IDs and Incorrect Answer Choices";
data _null_;
   file print;
   array ans[10] $ 1 Ans1-Ans10;
   infile 'c:\books\test scoring\data_errors.txt' pad;
   input @1  ID $9.
         @11 (Ans1-Ans10)($upcase1.);
   if notdigit(ID) and _n_ gt 0 then put "Invalid ID " ID "in line "
_n_;
   Do Item = 1 to 10;
      if missing(Ans[Item]) then
      put "Item #" Item "Left Blank by Student " ID;
      else if Ans[Item] not in ('A','B','C','D','E') then
         put "Invalid Answer for Student " ID "for Item #"
         Item "entered as " Ans[Item];
   end;
run;
```

Because your multiple-choice responses may be in upper- or lowercase, this program reads the student data and converts it to uppercase. This is accomplished by using the $UPCASE1. informat. You can use the NOTDIGIT function to test if the student ID contains any non-digit characters. This useful function searches every position of a character value (including trailing blanks) and returns the position of the first non-digit. If the character value only contains digits, the functions returns a 0. In SAS, any numeric value that is not 0 or missing is considered to be true. So, if the ID contains any non-digit characters, regardless of where that value is located in the ID, the function returns a number from 1 to 9 (since the ID variable in this example is nine digits) and is therefore considered true. Because you used a FILE PRINT statement, the PUT statement following the test for non-digit values writes to the output device (the Output window if you are using SAS Display Manager). The variable _N_ is a value created when you run a DATA step. It counts iterations of the DATA step. Because you are reading one line of input data with every iteration of the DATA step, the value of _N_ tells you what line of the text file contains an ID error.

Following the test for non-digit ID values, you test each of the 10 answer values to see if a value other than A through E is found. If so, you print out the ID, the item number, and the value.

This program assumes that the answer key is included in the file to be tested—it does not test for a valid ID in the first line of the file. If this is not the case, omit the test for _N_ greater than one in the test for non-digit values.

Output from Program 4.1

```
Checking for Invalid IDs and Incorrect Answer Choices
Item #2 Left Blank by Student 123456789
Invalid Answer for Student 444444444 for Item #5
entered as ?
Invalid ID 777x77777 in line 9
```

Checking for ID Errors

You tested for non-digit values in the previous program. In this program, you want to go one step further and determine if any of the IDs in the student test file are missing in the student name file (containing ID values and names). To demonstrate this, another text file called invalid_test.txt was created from the original sample_test.txt file with one ID changed so that it cannot be found in the student name file (student.txt). When you merge the test data file and the student name file (by ID), you can determine if there are any names in the test data file that are not found in the student name file, as demonstrated next:

Program 4.2: Checking for Student IDs in the Test File that Are Not in the Student Name File

```
*Identifying IDs in the test data that are not present
 in the student name file;
data student_test;
   infile 'c:\books\test scoring\invalid_test.txt' firstobs=2;
   input @1  ID $9.;
run;

proc sort data=student_test;
   by ID;
run;

data student_name;
   length Name $ 15 ID $ 9;
   infile 'c:\books\test scoring\student_name.txt' dsd;
   input Name ID;
run;

proc sort data=student_name;
   by ID;
run;

title "Listing of Student IDs not in the Student Data File";
data _null_;
   file print;
   merge student_name(in=in_student_name) student_test;
   by ID;
   if not in_student_name then put "ID " ID "not found in name file";
run;
```

You start out by reading the student data from the file. Because the first line of this file contains the answer key, you use the INFILE option FIRSTOBS= to tell the program to begin reading the text data file starting with the second line. You are only checking for invalid IDs, so that is the only variable that you need to read. Next, you sort this data set by ID since that is required in the merge that is to follow.

You may already have a permanent student name file, but in this example you are re-creating it in a DATA step. You also sort this data set by ID to prepare for the merge.

In the final DATA step, you merge the two SAS data sets by ID. The SAS data set option IN= creates a temporary variable (that you called in_student_name) that has a value of true if that data set is making a contribution to the merge and false if the observation is not making a contribution to the merge. If there is an ID in the student_test data set that is not in the student_name data set, the variable in_student_name will be false. You want to print out the ID for any student whose name is not included in the student name file. The output looks like this:

Output from Program 4.2

```
Listing of Student IDs not in the Student Data File
ID 434444444 not found in name file
```

Using "Fuzzy" Matching to Identify an Invalid ID

Two possible reasons why a student ID is missing from the name file are that the name was never entered or that the student marked it incorrectly on the answer sheet. This section shows you how to look for close matches between the incorrect ID and IDs in the student file using a method sometimes described as *fuzzy matching*. SAS has a spelling distance function (SPEDIS) that is used to match words (usually names) where they may not be spelled exactly the same. This same function can be used to identify closely matched IDs. Let's try to see if we can use this method to find a close match to ID 434444444. Here is the program:

Program 4.3: Attempting to Find a Close Match for IDs in the Test File and the Name File

```
*Attempting to find a close match for ID 434444444 in the student name
file;
data student_test;
   infile 'c:\books\test scoring\invalid_test.txt' firstobs=2;
   input @1  ID $9.;
run;

data student_name;
   length Name $ 15 ID $ 9;
   infile 'c:\books\test scoring\student_name.txt' dsd;
   input Name ID;
run;

title "Searching for a Close Match for ID 434444444";
proc sql;
   select test.ID as ID_from_Test,
          name.ID as ID_from_Name,
          Name
   from student_test as test,
        student_name as name
   where spedis(test.ID,name.ID) between 1 and 15;
quit;
```

Even though the IDs consist of digits and not letters, you can still use the spelling distance function (SPEDIS) to compare two IDs. The program starts by creating the student_test and student_name data sets, as in the previous program. To look for possible matches between the ID on the test and the IDs in the name file, you create a Cartesian product using PROC SQL. A Cartesian product is all possible combinations of observations in each file matched against every observation in the other file. As a simple example, if File One contained:

File One

```
111111111 Ron
222222222 Jim
```

And File Two contained:

File Two

```
111111110
222222222
333333333
```

The Cartesian product of File One and File Two would be:

Cartesian Product of File One and File Two

```
111111111 Ron 111111110
111111111 Ron 222222222
111111111 Ron 333333333
222222222 Jim 111111110
222222222 Jim 222222222
222222222 Jim 333333333
```

You can then look at each line in the Cartesian product data set to see if the two IDs are close. For those readers who enjoy the details, the SPEDIS function uses a spelling distance algorithm to compute a spelling distance between words. If the two words being compared are exactly the same, the spelling distance is 0. For each spelling mistake, the function assigns penalty points. Some spelling mistakes, such as getting the first letter wrong, result in a large penalty; other errors, such as transposing the position of two letters, result in a smaller error. Finally, after all the penalty points are computed, the spelling distance is computed as the sum of the penalty points as a percentage of the number of letters in the word you list as the first argument in the function. This makes sense since one letter wrong in a three-letter word is a rather bad spelling mistake, whereas one letter wrong in a 10-letter word is a minor error.

Now that you see how the SPEDIS function can compare IDs, let's see the result of running Program 4.3:

Output from Program 4.3

```
Searching for a Close Match for ID 434444444

ID_from_Test  ID_from_Name  Name
ƒƒƒƒƒƒƒƒƒƒƒƒƒƒƒƒƒƒƒƒƒƒƒƒƒƒƒƒƒƒƒƒƒƒƒƒƒƒƒƒƒƒƒƒ
434444444     444444444     Raul Heizer
```

Here you see a possible match for ID 434444444. You may want to experiment with spelling distance values in this program. If you pick a value too small, you may miss possible matches; if you pick a value too large, you may match IDs that do not belong to the same person.

Checking for and Eliminating Duplicate Records

There are a number of ways that duplicate records can occur. If you are using a scanner (with or without automatic feed), an answer sheet may cause a jam. When this happens, you usually scan the sheet a second time. Depending on the situation, the rescanning process may cause a duplicate record to be entered into your file. Another possible source of duplicate records results when your data is entered by more than one person and one or more answer sheets are entered multiple times.

The solution to this problem is very simple: When you use PROC SORT to sort your answer file, you can include an option to remove duplicate observations. To demonstrate this, two duplicate entries were added to the data_errors.txt file. This file was called duplicates.txt and is displayed next:

Listing of duplicates.txt

```
000000000 abcedaabed
123456789 a cdeaabed
111111111 abcedaabed
111111111 abcedaabed
222222222 cddabedeed
333333333 bbdddaaccd
444444444 abce?aabbb
555555555 eecedabbca
666666666 aecedaabed
888888888 bbccdeebed
777x77777 dbcaaaabed
888888888 bbccdeebed
999999999 cdcdadabed
```

The duplicate entries are printed in bold type. The following program detects and eliminates these duplicate observations:

Program 4.4: Program to Detect and Eliminate Duplicate Records in the Answer File

```
data check_for_dups;
   infile 'c:\books\test scoring\duplicates.txt' pad;
   input @1 ID $9.
         @11 (Ans1-Ans10)($upcase1.);
run;

proc sort data=check_for_dups noduprecs;
   by ID;
run;
```

The DATA step reads text data from the duplicates.txt data file. Although the first line of data is really the answer key, for purposes of detecting duplicates, you can treat it the same as student answers. The PROC SORT option NODUPRECS (no duplicate records) checks for duplicate records (that is, two records that are exact duplicates), keeps the first one, and deletes successive duplicates. It is a good idea to include this test before you attempt to score a test or perform item analysis. It is especially important to remove duplicate records if you plan to merge your test data with a file containing student names and IDs.

You can see in the listing of the SAS Log (below) that two duplicate records were detected and deleted.

SAS Log from Program 4.4

```
     proc sort data=check_for_dups noduprecs;
        by ID;
     run;

NOTE: There were 13 observations read from the data set
      WORK.CHECK_FOR_DUPS.
NOTE: 2 duplicate observations were deleted.
NOTE: The data set WORK.CHECK_FOR_DUPS has 11 observations and 11
      variables.
NOTE: PROCEDURE SORT used (Total process time):
      real time           0.04 seconds
      cpu time            0.01 seconds
```

The note in the SAS Log tells you that 13 observations were read, 2 duplicate observations were deleted, and data set CHECK_FOR_DUPS has 11 observations.

Conclusion

You should routinely perform data validity checks before attempting to run any of the item analysis programs described in this book. A possible exception to this recommendation might be considered if you are using a scanner with sophisticated software that performs data integrity checks as the tests are scanned. This is not the case in most scanners on the market.

A SAS macro is included in Chapter 11 to automate the task performed by Program 4.1.

Chapter 5: Classical Item Analysis

Introduction

This chapter investigates some traditional methods of determining how well items are performing on a multiple-choice (or true/false) test. A later chapter covers advances in item response theory.

Point-Biserial Correlation Coefficient

One of the most popular methods for determining how well an item is performing on a test is called the *point-biserial correlation coefficient*. Computationally, it is equivalent to a Pearson correlation between an item response (correct=1, incorrect=0) and the test score for each student. The simplest way for SAS to produce point-biserial coefficients is by using PROC CORR. Later in this chapter, you will see a program that computes this value in a DATA step and displays it in a more compact form than PROC CORR. The following program produces correlations for the first 10 items in the statistics test described in Chapter 2.

Program 5.1: Computing Correlations Between Item Scores and Raw Scores

```
title "Computing Point-Biserial Correlations";
proc corr data=score nosimple;
   var Score1-Score10;
   with Raw;
run;
```

When you supply PROC CORR with a VAR statement and a WITH statement, it computes correlations between every variable listed on the VAR statement and every variable listed on the WITH statement. If you only supply a VAR statement, PROC CORR computes a correlation matrix—the correlation of every variable in the list with every other variable in the list.

Output from Program 5.1:

Computing Point-Biserial Correlations

The CORR Procedure

1 With Variables:	Raw
10 Variables:	Score1 Score2 Score3 Score4 Score5 Score6 Score7 Score8 Score9 Score10

Pearson Correlation Coefficients, N = 137 Prob > |r| under H0: Rho=0										
	Score1	**Score2**	**Score3**	**Score4**	**Score5**	**Score6**	**Score7**	**Score8**	**Score9**	**Score10**
Raw Raw score	0.48830 <.0001	0.29773 0.0004	0.32225 0.0001	0.43235 <.0001	0.35183 <.0001	0.20284 0.0174	0.41984 <.0001	0.06532 0.4483	0.27607 0.0011	0.38473 <.0001

The top number in each box is the correlation between the item and the raw test score, referred to as a *point-biserial correlation coefficient*. The number below this is the *p*-value (significance level). How do you interpret this correlation? One definition of a "good" item is one where good students (those who did well on the test) get the item correct more often than students who do poorly on the test. This condition results in a positive point-biserial coefficient. Since the distribution of test scores is mostly continuous and the item scores are dichotomous (0 or 1), this correlation is usually not as large as one between two continuous variables. What does it tell you if the point-biserial correlation is close to 0? It means the "good" and "poor" students are doing equally well answering the item, meaning that the item is not helping to discriminate between good and poor students. What about negative coefficients? That situation usually results from several possible causes: One possibility is that there is a mistake in the answer key—good students are getting it wrong quite frequently (they are actually choosing the correct answer, but it doesn't match the answer key) and poor students are guessing their answers and getting the item right by chance. Another possibility is a poorly written item that good students are "reading into" and poor students are not. For example, there might be an item that uses absolutes such as "always" or "never" and the better students can think of a rare exception and do not choose the answer you expect. A third possibility is that the item is measuring something other than, or in addition to, what you are interested in. For example, on a math test, you might have a word problem where the math is not all that challenging, but the language is a bit subtle. Thus, students with better verbal skills are getting the item right as opposed to those with better math skills. Sometimes an answer that you thought was incorrect might be appealing to the better students, and upon reflection, you conclude, "Yeah, that *could* be seen as correct." There are other possibilities as well, which we will explore later.

Making a More Attractive Report

Although the information in the previous output has all the values you need, it is hard to read, especially when there are a lot of items on the test. The program shown next produces a much better display of this information.

Program 5.2: Producing a More Attractive Report

```
proc corr data=score nosimple noprint
          outp=corrout;
   var Score1-Score10;
   with Raw;
run;
```

The first step is to have PROC CORR compute the correlations and place them in a SAS data set. To do this, you use an OUTP= procedure option. This places information about the selected variables, such as the correlation coefficients and the means, into the data set you specify with this option. The NOPRINT option is an instruction to PROC CORR that you do not want any printed output, just the data set. Here is a listing of data set CORROUT:

Listing of Data Set CORROUT

TYPE	_NAME_	Score1	Score2	Score3	Score4	Score5	Score6	Score7	Scor
MEAN		0.620	0.818	0.934	0.460	0.854	0.956	0.635	0.83
STD		0.487	0.388	0.249	0.500	0.354	0.205	0.483	0.375
N		137.000	137.000	137.000	137.000	137.000	137.000	137.000	137.000
CORR	Raw	0.488	0.298	0.322	0.432	0.352	0.203	0.420	0.065

Note: The last few columns were deleted to allow the table to fit better on the page.

This data set contains the mean, the standard deviation, N, and a correlation for each of the Score variables. The SAS created variable, _TYPE_, identifies which of these values you are looking at—the variable _NAME_ identifies the name of the WITH variable. In this example, you are only interested in the correlation coefficients. An easy way to subset the CORROUT data so that it only contains correlation coefficients is with a WHERE= data set option following the data set name. The following program is identical to Program 5.2 with the addition of the WHERE= data set option. Here is the modified program, followed by a listing of the CORROUT data set:

Program 5.3: Adding a WHERE= Data Set Option to Subset the SAS Data Set

```
proc corr data=score nosimple noprint
          outp=corrout(where=(_type_='CORR'));
   var Score1-Score10;
   with Raw;
run;
```

By using the WHERE= data set option, you now have only the correlation data in the output data set.

Listing of Data Set CORROUT (Created by Program 5.3)

TYPE	_NAME_	Score1	Score2	Score3	Score4	Score5	Score6	Score7	Score8
CORR	Raw	0.48830	0.29773	0.32225	0.43235	0.35183	0.20284	0.41984	0.0655

Note: The last few columns were deleted to allow the table to fit better on the page.

Program 5.3 is a good example of programming efficiently—using a WHERE= data set option when the data set is being created rather than writing a separate data set to create the subset.

The Next Step: Restructuring the Data Set

The next step in creating your report is to restructure (transpose) the data set above into one with one observation per item. Although you could use PROC TRANSPOSE to do this, an easier (at least to these authors) method is to use a DATA step, as follows:

Program 5.4: Restructuring the Correlation Data Set to Create a Report

```
data corr;
   set corrout;
   array Score[10];
   do Item=1 to 10;
      Corr = Score[Item];
      output;
   end;
   keep Item Corr;
run;
```

You read in one observation from data set CORROUT and, inside the DO loop, you write out one observation for each of the 10 correlations. The newly created data set (CORR) looks like this:

Listing of Data Set CORR

Item	Corr
1	0.48830
2	0.29773
3	0.32225
4	0.43235
5	0.35183
6	0.20284
7	0.41984
8	0.06532
9	0.27607
10	0.38473

You now have the 10 point-biserial coefficients in a more compact, easier-to-read format. In a later section, this information is combined with item frequencies in a single table.

Displaying the Mean Score of the Students Who Chose Each of the Multiple Choices

One interesting way to help you determine how well your items are performing is to compute the mean score for all of the students choosing each of the multiple-choice items.

To demonstrate this, let's start out with test data from a real test (a biostatistics test taken by students at a medical school in New Jersey—all the IDs have been replaced with random digits). A listing of the first 10 students with responses to the first eight items on the test are shown below:

Listing of the First Eight Items for 10 Students

ID	Ans1	Ans2	Ans3	Ans4	Ans5	Ans6	Ans7	Ans8	Raw	Percent
203579875	D*	C*	E*	B*	C*	B*	A*	B*	47	83.9286
116841443	D*	C*	E*	A	C*	B*	A*	B*	41	73.2143
176786926	D*	C*	B	B*	C*	B*	B	D	45	80.3571
011413555	D*	C*	E*	A	C*	B*	E	B*	39	69.6429
051502147	E	E	E*	B*	C*	B*	A*	B*	41	73.2143
069456918	D*	C*	E*	B*	C*	B*	A*	B*	46	82.1429
743222519	D*	C*	E*	B*	C*	B*	B	B*	33	58.9286
177458805	E	C*	E*	B*	C*	B*	A*	B*	42	75.0000

In this data set, the correct answer choice includes the letter (A through E) followed by an asterisk (as described in Chapter 3). You can run PROC FREQ on the answer variables like this. (Note: Frequencies for only the first four items were requested to limit the size of the output.)

Program 5.5: Using PROC FREQ to Determine Answer Frequencies

```
title "Frequencies for the First 4 Items on the Biostatistics Test";
proc freq data=score;
   tables Ans1-Ans4 / nocum;
run;
```

You now have answer frequencies with the correct answer to each item displayed with an asterisk.

Answer Frequencies for the First Four Items on the Biostatistics Test

Frequencies for the First 4 Items on the Biostatistics Test

The FREQ Procedure

Ans1	Frequency	Percent
A	1	0.73
B	2	1.46
D*	85	62.04
E	49	35.77

Ans2	Frequency	Percent
A	2	1.47
B	7	5.15
C*	112	82.35
D	1	0.74
E	14	10.29
Frequency Missing = 1		

Ans3	Frequency	Percent
B	7	5.11
C	2	1.46
E*	128	93.43

Ans4	Frequency	Percent
A	23	16.91
B*	63	46.32
C	24	17.65
D	17	12.50
E	9	6.62
Frequency Missing = 1		

The next step is to compute the mean score for all students who chose each of the multiple-choice answers. To do this, you must first restructure the SCORE data set so that you have one observation per student per item. The program to accomplish this restructuring is displayed next:

Program 5.6: Restructuring the Score Data Set with One Observation per Student per Question

```
data restructure;
   set score;
   array Ans[*] $ 2 Ans1-Ans10;
   do Item=1 to 10;
      Choice=Ans[Item];
      output;
   end;
   keep Item Choice Percent;
run;
```

For each observation in the SCORE data set, you output 10 observations in the RESTRUCTURE data set, one observation for each item. Here are the first few observations in the RESTRUCTURE data set:

First 20 Observations in Data Set RESTRUCTURE

Percent	Item	Choice
83.9286	1	D*
83.9286	2	C*
83.9286	3	E*
83.9286	4	B*
83.9286	5	C*
83.9286	6	B*
83.9286	7	A*
83.9286	8	B*
83.9286	9	C*
83.9286	10	E*
73.2143	1	D*
73.2143	2	C*
73.2143	3	E*
73.2143	4	A
73.2143	5	C*
73.2143	6	B*
73.2143	7	A*
73.2143	8	B*
73.2143	9	A
73.2143	10	D

Using this data set, you can now compute the mean score for all students who chose each of the answers. You can use PROC MEANS to do this, using the item number and the answer choice as class variables. That is, you want to compute the mean value of Percent for each combination of Item and Class. Here is the code:

Program 5.7: Using PROC MEANS to Compute the Mean Percent for Each Combination of Item and Choice

```
proc means data=restructure mean std maxdec=2;
   class Item Choice;
   var Percent;
run;
```

The resulting output is listed next (only the first few values are displayed here):

Output from Program 5.7

The MEANS Procedure

Analysis Variable : Percent Percent score				
Item	**Choice**	**N Obs**	**Mean**	**Std Dev**
1	A	1	46.43	.
	B	2	69.64	17.68
	D*	85	77.10	7.50
	E	49	66.55	11.74
2	A	2	55.36	5.05
	B	7	70.92	5.81
	C*	112	74.51	10.51
	D	1	64.29	.
	E	14	66.84	9.75
3	B	7	63.27	15.18
	C	2	48.21	2.53
	E*	128	73.91	9.91
4	A	23	68.63	11.37
	B*	63	78.03	8.86
	C	24	68.97	11.59
	D	17	69.12	8.01
	E	9	67.46	11.39
5	A	8	61.83	15.97
	B	9	61.71	12.15
	C*	117	74.56	9.42
	D	2	80.36	2.53

In each of the first four items, the mean score of all students choosing the right answer is higher than for any other choice. However, you can see that students choosing B for item two have a mean score (70.92) almost as high as that for students choosing the correct answer (C average score = 74.51). You might want to examine choice D to see if you want to make changes.

Combining the Mean Score per Answer Choice with Frequency Counts

To make the above display even more useful, you can combine the mean score per answer choice with the answer frequencies. The best way to do this is with PROC TABULATE. This SAS procedure can combine statistical and frequency data in a single table. The PROC TABULATE statements are shown in the next program:

Program 5.8: Using PROC TABULATE to Combine Means Scores and Answer Frequencies

```
proc format;
   picture pct low-<0=' ' 0-high='009.9%';
run;

title "Displaying the Student Mean Score for Each Answer Choice";
proc tabulate data=restructure;
   class Item Choice;
   var Percent;
   table Item*Choice,
   Percent=' '*(pctn<Choice>*f=pct. mean*f=pct.
   std*f=10.2)  / rts=20 misstext=' ';
   keylabel all  = 'Total'
            mean = 'Mean Score' pctn='Freq'
            std  = 'Standard Deviation';
run;
```

The picture format in this program prints percentages to a tenth of a percent and adds the percent sign to the value. Two class variables, Item and Choice, are used to show statistics and counts for each combination of Item and Choice (the same as with PROC MEANS, described earlier). The VAR statement lists all the variables for which you want to compute statistics. Because you want to see the mean percentage score for each item and choice, you list Percent on the VAR statement. Finally, the TABLES statement defines the rows and columns in the table. We will skip some details and only indicate that the rows of the table contain values of Choice nested within Item and the columns of the table contain percentages (the keyword PCT does this), means, and standard deviations. The remainder of the statements select formats for the various values as well as labels that you want to associate with each statistic. The resulting table is shown next:

Output from Program 5.8 (Partial Listing)

Displaying the Student Mean Score for Each Answer Choice

Item	Choice	Freq	Mean Score	Standard Deviation
1	A	0.7%	46.4%	
	B	1.4%	69.6%	17.68
	D*	62.0%	77.1%	7.50
	E	35.7%	66.5%	11.74
2	A	1.4%	55.3%	5.05
	B	5.1%	70.9%	5.81
	C*	82.3%	74.5%	10.51
	D	0.7%	64.2%	
	E	10.2%	66.8%	9.75
3	B	5.1%	63.2%	15.18
	C	1.4%	48.2%	2.53
	E*	93.4%	73.9%	9.91
4	A	16.9%	68.6%	11.37
	B*	46.3%	78.0%	8.86
	C	17.6%	68.9%	11.59
	D	12.5%	69.1%	8.01
	E	6.6%	67.4%	11.39
[illegible]	A	5.8%	[illegible]	15.97

You now have the answer frequencies and the student mean scores for each answer choice for each item on the test in a single table. Take a look at item one. Notice that the students who answered this item correctly (answer D) had a mean test score of 77.10, which is higher than the mean test score for any of the incorrect responses.

Later sections of this chapter will build on this and add even more item information in a single table.

Computing the Proportion Correct by Quartile

Besides inspecting the point-biserial correlations (a single number), you will gain more insight into how an item is performing by dividing the class into quantiles. The number of quantiles will depend on how many students took the test—if you have a relatively small sample size, you may only want three or four quantiles. For much larger samples, you may want as many as six. Once you have divided the class into quantiles, you can then compute the proportion of students answering an item correctly in each quantile. What you hope to see is the proportion correct increasing, going from the lowest quantile to the highest. You can examine the proportions correct by quantile in tabular form or produce a graphical plot of this information. Such a plot is called an *item characteristic curve.*

In this example, you are going to use a 56-item statistics test administered to 137 students. For this sample size, dividing the class into quartiles (four groups) works well. You can use the SAS RANK procedure to divide a data set into any number of groups. Here is the first step:

Program 5.9: Dividing the Group into Quartiles

```
*Dividing the group into quartiles;
proc rank data=score(keep=Raw Score1-Score10) groups=4 out=quartiles;
   var Raw;
   ranks Quartile;
run;
```

You use the GROUPS= option to tell PROC RANK to divide the data set into four groups, based on the value of the variable Raw. The RANKS statement names the variable that will contain the group numbers. For some completely unknown reason, when you ask PROC RANK to create groups, it numbers the groups starting from 0. Because you are requesting four groups, the variable Quartile will have values from 0 to 3, as shown below:

Output from Program 5.9 (First 10 Observations)

Listing of Data Set QUARTILES

ID	Raw	Quartile	Score1	Score2	Score3	Score4	Score5
203579875	47	3	1	1	1	1	1
116841443	41	1	1	1	1	0	1
176786926	45	2	1	1	0	1	1
011413555	39	1	1	1	1	0	1
051502147	41	1	0	0	1	1	1
069456918	46	3	1	1	1	1	1
743222519	33	0	1	1	1	1	1
177458805	42	2	0	1	1	1	1
927950674	36	0	1	0	0	0	1
181246859	35	0	0	1	1	0	1

You can now compute the mean score for each quartile, like this:

Program 5.10: Computing the Mean Scores by Quartile

```
title "Mean Scores by Quartile - First 10 Items";
proc means data=quartiles mean maxdec=2;
   class Quartile;
   Var Score1-Score10;
run;
```

Here is the output:

Output from Program 5.10

Mean Scores by Quartile - First 10 Items

The MEANS Procedure

Rank for Variable Raw	N Obs	Variable	Mean
0	31	Score1	0.23
		Score2	0.61
		Score3	0.81
		Score4	0.16
		Score5	0.65
		Score6	0.87
		Score7	0.32
		Score8	0.77
		Score9	0.81
		Score10	0.71
1	35	Score1	0.69
		Score2	0.71
		Score3	0.94
		Score4	0.34
		Score5	0.91
		Score6	0.97
		Score7	0.63
		Score8	0.83
		Score9	0.80
		Score10	0.91
2	40	Score1	0.68
		Score2	0.98
		Score3	0.98
		Score4	0.53
		Score5	0.90
		Score6	1.00
		Score7	0.65
		Score8	0.83
		Score9	0.95
		Score10	0.98
3	31	Score1	0.87
		Score2	0.94
		Score3	1.00
		Score4	0.81
		Score5	0.94
		Score6	0.97
		Score7	0.94
		Score8	0.90
		Score9	1.00
		Score10	0.94

Although this listing contains the mean score for each of the four quartiles for the first 10 items on the test, the layout is inconvenient. You would rather see each item in order, with the percent correct by quartile displayed on a single line. The program to restructure and combine the answer frequencies for each of the test items, the item difficulty, the point-biserial coefficient, and the proportion correct by quartile is the subject of the next section.

Combining All the Item Statistics in a Single Table

In order to demonstrate this final program, the data from the 56-item statistics test is used. However, only the first 10 items on the test are analyzed (to reduce the size of the generated reports). The scores used to divide the class into quartiles are based on all 56 items.

The programming to combine all of the item statistics into a single table takes a number of steps and is not for the faint of heart. You can skip right to the output and its interpretation if you wish. For those readers who want to understand the programming details, the program contains callouts that link to descriptions following the program.

Program 5.11: Combining All the Item Statistics in a Single Table

```
%let Nitems=56; ❶
data score;
   infile 'c:\books\test scoring\stat_test.txt' pad;
   array Ans[&Nitems] $ 1 Ans1-Ans&Nitems;      *Student Answers;
   array Key[&Nitems] $ 1 Key1-Key&Nitems;      *Answer Key;
   array Score[&Nitems] 3 Score1-Score&Nitems; *1=right,0=wrong;
   retain Key1-Key&Nitems;
   if _n_=1 then input @11 (Key1-Key&Nitems)($1.);
   input @1  ID $9.
         @11 (Ans1-Ans&Nitems)($1.);
   do Item = 1 to &Nitems;
      Score[Item] = Key[Item] eq Ans[Item];
   end;
   Raw=sum (of Score1-Score&Nitems);
   Percent=100*Raw / &Nitems;
   keep Ans1-Ans&Nitems Key1-Key&Nitems ID Raw Percent Score1-
Score&Nitems;
   label ID      = 'Student ID'
         Raw     = 'Raw score'
         Percent = 'Percent score';
run;

*Divide the group into quartiles; ❷
proc rank data=score groups=4 out=quartiles;
   var Raw;
   ranks Quartile;
run;
```

```
*Restructure the data set so that you have one item per
 observation; ❸

data tab;
   set quartiles;
   length Choice $ 1 Item_Key $ 5;
   array Score[10] Score1-Score10;
   array ans[10] $ 1 Ans1-Ans10;
   array key[10] $ 1 Key1-Key10;
   Quartile = Quartile + 1;
   do Item=1 to 10;
      Item_Key = cat(right(put(Item,3.))," ",Key[Item]);
      Correct=Score[Item];
      Choice=Ans[Item];
      output;
   end;
   keep Item Item_Key Quartile Correct Choice;
run;

*Sort by item number; ❹
proc sort data=tab;
   by Item;
run;

*Compute correlation coefficients; ❺
proc corr data=score nosimple noprint
          outp=corrout(where=(_type_='CORR'));
   var Score1-Score10;
   with Raw;
run;

*Restructure correlation data set; ❻
data corr;
   set corrout;
   array Score[10];
   do Item=1 to 10;
      Corr = Score[Item];
      output;
   end;
   keep Item Corr;
run;

*Combine correlations and quartile information; ❼
data both;
   merge corr tab;
   by Item;
run;
```

```
*Print out final table;
proc tabulate format=7.2 data=both order=internal noseps;
   title "Item Statistics";
   label Quartile = 'Quartile'
         Choice   = 'Choices';
   class Item_Key Quartile Choice;
   var Correct Corr;
   table Item_Key = 'Num Key'*f=6. ,
      Choice*(pctn<Choice>)*f=3.
      Correct=' '*mean='Diff.'*f=percent5.2
      Corr=' '*mean='Corr.'*f=5.2
      Correct=' '*Quartile*mean='Prop. Correct'*f=percent7.2/
      rts=8;
   keylabel pctn='%' ;
run;
```

❶ You start out by scoring the test in the usual way. Instead of hard coding the number of items on the test, you use the macro variable &Nitems and assign the number of items (56) with a %LET statement. Using %LET is another way of assigning a value to a macro variable. There is one other small difference in this program compared to the scoring programs displayed previously: The variables in the Score array are stored in 3 bytes (notice the 3 before the list of variables in this array). Since the values of the Score variables are 0s and 1s, you do not need to store them in 8 bytes (the default storage length for SAS numeric values). Three bytes is the minimum length allowed by SAS.

❷ You use PROC RANK to create a variable (that you call Quartile) that represents quartiles of the Raw score. Remember that the values of Quartile range from 0 to 3. Below are the first few observations from data set QUARTILES (with selected variables displayed):

First Few Observations in Data Set Quartiles (Selected Variables Displayed)

Listing of Data Set QUARTILES

Raw	Quartile	Ans1	Ans2	Key1	Key2	Score1	Score2
47	3	D	C	D	C	1	1
41	1	D	C	D	C	1	1
45	2	D	C	D	C	1	1
39	1	D	C	D	C	1	1
41	1	E	E	D	C	0	0
46	3	D	C	D	C	1	1
33	0	D	C	D	C	1	1
42	2	E	C	D	C	0	1
36	0	D	E	D	C	1	0

❸ You need to restructure this data set so that there is one observation per test item. That way it can be combined with the correlation data and later displayed in a table with one item per row. In this DATA step, you accomplish several goals. First, you add 1 to Quartile so that the values now range from 1 to 4 (instead of 0 to 3). Next, you create a new variable (Item_Key) that puts together (*concatenates*, in computer jargon) the item number and the answer key. Each of the item scores (the 0s and 1s) is assigned to the variable Correct. Finally, each answer choice is assigned to a variable you call Choice.

❹ You sort the TAB data set by item (so that you can combine it with the correlation data). Here are the first 20 observations from the sorted data set. Looking at this listing should help you understand step 3:

First 20 Observations from Data Set TAB after Sorting

Listing of Data Set TAB

Quartile	Choice	Item_Key	Item	Correct
4	D	1D	1	1
2	D	1D	1	1
3	D	1D	1	1
2	D	1D	1	1
2	E	1D	1	0
4	D	1D	1	1
1	D	1D	1	1
3	E	1D	1	0
1	D	1D	1	1
1	E	1D	1	0
4	D	1D	1	1
4	E	1D	1	0
3	D	1D	1	1
3	D	1D	1	1
4	E	1D	1	0
1	E	1D	1	0
3	D	1D	1	1
3	D	1D	1	1
4	D	1D	1	1
4	D	1D	1	1

The mean of the variable Correct for Item would be the proportion of the entire class that answered the item correctly. If you compute the mean for each of the four quartiles of the class, you have the proportion correct by quartile, one of the values you want to display in the final table.

❺ You use PROC CORR to compute the point-biserial correlations for each of the items and place these correlations in a data set called CORROUT. Here is the listing of this data set:

Listing of Data Set CORROUT

TYPE	_NAME_	Score1	Score2	Score3	Score4	Score5	Score6	Score7	Score8	Score9	Score10
CORR	Raw	0.48830	0.29773	0.32225	0.43235	0.35183	0.20284	0.41984	0.065315	0.27607	0.38473

❻ You now restructure the correlation data set so that there is one observation per item. Here is the listing:

Listing of Data Set CORR

Item	Corr
1	0.48830
2	0.29773
3	0.32225
4	0.43235
5	0.35183
6	0.20284
7	0.41984
8	0.06532
9	0.27607
10	0.38473

❼ You can now combine the data from the two data sets (TAB and CORR) since they are now in Item order.

❽ You use PROC TABULATE to compute the mean value of Correct for each of the quartiles and display all the statistics for each item. PROC TABULATE allows you to format each of the cells in the table. After all this work, here is the final table:

Item Statistics

	Choices							Quartile			
	A	B	C	D	E			1	2	3	4
	%	%	%	%	%	Diff.	Corr.	Prop. Correct	Prop. Correct	Prop. Correct	Prop. Correct
Num Key											
1 D	1	1	.	62	36	62%	0.49	22.6%	68.6%	67.5%	87.1%
2 C	1	5	82	1	10	82%	0.30	63.3%	71.4%	97.5%	93.5%
3 E	.	5	1	.	93	93%	0.32	80.6%	94.3%	97.5%	100%
4 B	17	46	18	13	7	46%	0.43	16.1%	35.3%	52.5%	80.6%
5 C	6	7	85	1	1	85%	0.35	64.5%	91.4%	90.0%	93.5%
6 B	1	96	3	.	.	96%	0.20	87.1%	97.1%	100%	96.8%
7 A	64	21	6	1	8	64%	0.42	32.3%	62.9%	65.0%	93.5%
8 B	7	84	.	4	5	84%	0.07	77.4%	82.9%	82.5%	93.3%
9 C	9	.	89	1	1	89%	0.28	80.6%	80.0%	95.0%	100%
10 E	.	3	4	4	89	89%	0.38	71.0%	91.4%	97.5%	93.5%

Yes, that was a lot of work, but you can now see all of the item information (the percent of the class choosing each of the answer choices, the item difficulty, the point-biserial coefficient, and the proportion correct by quartile) in a single table. SAS macros (pre-packaged programs) to accomplish all the tasks described in this book can be found in Chapter 11.

Interpreting the Item Statistics

The definition of a "good" item depends somewhat on why you are testing people in the first place. If the test is designed to rank students by ability in a class, you would like items with high point-biserial correlations and an increasing value of the percent-by-quartile statistic. For example, take a look at item 4. Notice that all of the answer choices have been selected. This indicates that there are no obvious wrong answers that all the students can reject. Next, notice that this is a fairly difficult item—only 46% of the students answered this item correctly. As a teacher, you may find this disappointing, but as a psychometrician, you are pleased that this item has a fairly high point-biserial correlation (.43) and the proportion of the students answering the item correctly increases from 16.1% to 80.6% over the four quartiles.

Let's look at another item. Most students answered item 3 correctly (difficulty = 93%). Items that are very easy or very hard are not as useful in discriminating student ability as other items. Item 3, although very easy, still shows an increase in the proportion by quartile but, because this increase is not as dramatic as item 4, the point-biserial correlation is a bit lower. Suppose you had an item that every student answered correctly. Obviously, this item would not be useful in discriminating good students from poor students (the point-biserial correlation would be 0). If your goal is to determine whether students understand certain course objectives, you may decide that it is OK to keep items that almost all students answer correctly.

Conclusion

Each of us has taken tests with poorly written items. We see a question that uses words like "every" or "never." You can think of one or two really rare exceptions and wonder: Is the teacher trying to trick me or am I reading into the question? Not only are poorly written items frustrating to the test taker, but they also reduce the test's reliability. Using the methods and programs described in this chapter is a good first step in identifying items that need improvement.

Chapter 6: Adding Special Features to the Scoring Program

Introduction

This chapter shows you how to add alternate correct answers to test items, how to rescore a test after deleting items, and how to score a test that has alternate versions. The first topic is of use when you have an item that you do not want to delete from the test but, after reviewing the item, you believe that you should accept an alternate answer to that item. Deleting items and rescoring a test is useful if your item analysis identifies one or more poor items. Finally, having alternate versions of a test (the same items but in different order) is a useful technique to discourage cheating.

Modifying the Scoring Program to Accept Alternate Correct Answers

At the Robert Wood Johnson Medical School, where one of the authors worked, a number of the professors requested that the test scoring program have the ability to accept alternate correct answers. The program discussed next demonstrates one way to allow a scoring program to accept alternate correct answers.

There are several ways to supply alternate answers to your test items. One way is to create a separate data set with each observation containing an item number and the alternate correct answers. Another way, demonstrated here, is to supply the answer key (including the alternate correct values) as the first line of the student data file. For this example, you want to accept answers b and c for item 2 and answers a, c, and d for item 6. Here is a listing of the data file:

Listing of File alternate.txt

```
a bd c e d acd a b e d
123456789 abcdeaabed
111111111 abcedaabed
222222222 cddabedeed
333333333 bbdddaaccd
444444444 abcedaabbb
555555555 eecedabbca
666666666 aecedaabed
777777777 dbcaaaabed
888888888 bbccdeebed
999999999 cdcdadabed
```

The first line of this file contains the correct answers to each of the test items—the remaining lines contain the student IDs and answers. The following program demonstrates how to include the alternate answers in the scoring process:

Program 6.1: Modifying the Scoring Program to Accept Alternate Answers

```
*Program to demonstrate how to process multiple correct
 answers;
data alternate_correct;
   infile 'c:\books\test scoring\alternate.txt' pad;
   array Ans[10] $ 1 Ans1-Ans10;    ***student answers;
   array Key[10] $ 4 Key1-Key10;    ***answer key;
   array Score[10] Score1-Score10; ***score array 1=right,0=wrong;
   retain Key1-Key10;
   if _n_ = 1 then input (Key1-Key10)(: $4.);
   input @1  ID $9.
         @11 (Ans1-Ans10)($1.);
   do Item=1 to 10;
      if findc(Ans[Item],Key[Item],'i') then Score[Item] = 1;
      else score[Item] = 0;
   end;
```

```
   Raw=sum(of Score1-Score10);
   Percent=100*Raw / 10;
   keep Ans1-Ans10 Key1-Key10 ID Raw Percent;
   label ID      = 'Student ID'
         Raw     = 'Raw Score'
         Percent = 'Percent Score';
run;
```

If you prefer using commas instead of blanks to separate the answer key values, add the DSD option to the INFILE statement as demonstrated in Chapter 2 (Program 2.3).

Instead of storing each answer key value in a single byte, you allocate four bytes for each key value (although it would be unusual to have more than one alternate correct answer). The INPUT statement that reads the 10 key values reads up to four values and stops when it reaches a delimiter (a blank in this example). The colon in the informat list (the list in parentheses following the list containing the 10 Key variables) is the instruction not to read past the delimiter.

Comparing the answer key to the student answer is done using the FINDC function. This function searches the first argument (the student answer) for any one of the values in the corresponding answer key. The third argument (the 'i') is a modifier that tells the function to ignore case. If the student answer matches any one of the answer key values, the Score variable is set to 1; otherwise, it is set to 0. Output from this program is shown below:

Output from Program 6.1

Listing of ALTERNATE_CORRECT

Student ID	Raw Score	Percent Score
123456789	8	80
111111111	10	100
222222222	3	30
333333333	5	50
444444444	8	80
555555555	5	50
666666666	9	90
777777777	7	70
888888888	6	60
999999999	7	70

As a quick check to verify that students choosing alternate answers are getting credit, let's take a look at the responses for student 222222222. Responses for this student to the 10 items were:

```
222222222 1=c, 2=d, 3=d, 4=a, 5=b, 6=e, 7=d, 8=e, 9=e, 10=d
```

and the answer key was:

```
1=a, 2=bd, 3=c, 4=e, 5=d, 6=acd, 7=a, 8=b, 9=e, 10=d
```

Therefore, this student answered three items correctly (getting credit for items 2, 9, and 10) and received a raw score of three (Note: This student needs to study more!).

Deleting Items and Rescoring the Test

Once you have inspected your item statistics, you may decide that you want to delete some items and rescore the test. Doing this can increase the overall reliability of the test. To demonstrate this process, the statistics test used in previous examples was scored again with the answer key modified to contain an incorrect choice for the first five items on the test. This obviously affects the item statistics for these items, as well as the overall test reliability. The following output displays the item statistics (for the first few items) on this modified test:

Item Analysis for the First Few Items on the Statistics Test with a Modified Answer Key

Item Statistics

	Choices							Quartile			
	A	B	C	D	E			1	2	3	4
	%	%	%	%	%	Diff.	Corr.	Prop. Correct	Prop. Correct	Prop. Correct	Prop. Correct
# Key											
1 A	1	1	.	62	36	.7%	-0.22	3.33%	0.00%	0.00%	0.00%
2 B	1	5	82	1	10	5%	0.01	3.45%	9.52%	2.70%	3.57%
3 A	.	5	1	.	93	.0%	.	0.00%	0.00%	0.00%	0.00%
4 C	17	46	18	13	7	18%	-0.09	20.7%	19.0%	16.2%	14.3%
5 D	6	7	85	1	1	1%	0.12	0.00%	0.00%	0.00%	7.14%
6 B	1	96	3	.	.	96%	0.18	86.7%	97.6%	100%	96.4%
7 A	64	21	6	1	8	64%	0.41	36.7%	57.1%	70.3%	92.9%
8 B	7	84	.	4	5	84%	0.04	80.0%	85.7%	78.4%	92.6%
9 C	9	.	89	1	1	89%	0.31	76.7%	81.0%	100%	100%
10 E	.	3	4	4	89	89%	0.38	73.3%	90.5%	94.6%	96.4%
1[illegible] A	[illegible]	[illegible]	[illegible]	1	6	8[illegible]	[illegible]	[illegible]	[illegible]	[illegible]	[illegible]

Notice that the item statistics for the first five items are quite poor.

The KR-20 reliability for the modified test (with the incorrect answer key) is .752, compared to a value of .798 on the original test (with the correct answer key). This demonstrates that including items that have low point-biserial coefficients may lower the test reliability.

You may decide to rescore this test with the first five items deleted. The following program demonstrates how to do this.

The short program that follows rescores the statistics test with the first five items deleted.

Program 6.2: Deleting Items and Rescoring the Test

```
%let List = 1 2 3 4 5;

data temp;
   set wrong_ans_stat(keep=ID Score1-Score56);
   retain Num;
   *List is the list of items to delete,
    Num is the number of items on the test after deletion;
   array Score[56] Score1-Score56;
   Num = 56;
   do Item = 1 to 56;
      Raw + Score[Item]*(Item not in (&List));
      if Item in(&List) then Num = Num - 1;
   end;
   Percent = 100*Raw/Num;
run;
```

The list of items to delete is assigned to a macro variable (&List) with a %LET statement. This method of assignment makes it easier to rerun the program with a different set of items to delete. It also makes it easier to convert the program to a macro.

The variable Num is the number of items on the original test minus the number of items you want to delete. The DO loop performs two functions. First, it sums the Score variables (1=correct, 0=incorrect) for all the items not in the delete list. The statement that performs the scoring combines a logical statement (`Item not in &list`) with an arithmetic computation (multiplying it by `Score[Item]`). This is how that statement works: If an item number is not in the delete list (&List), then the logical statement is true and SAS treats a true value as a 1, when combined with an arithmetic calculation. When an item is in the delete list, the logical statement is false. Since a false value is equal to 0, you multiply the score value (either a 0 or 1) by 0, thus ignoring deleted items in the scoring process. The second function of the DO loop computes the number of non-deleted items. You do this by decrementing the value of Num by one for each item in the delete list. The percent score is 100 times the raw score divided by the number of non-deleted items.

When you run this program with the first five items marked for deletion, the computed value of KR-20 is .779, an improvement over the value with the five items included (.752).

A macro version of this program is included in Chapter 11.

Analyzing Tests with Multiple Versions (with Correspondence Information in a Text File)

In testing conditions where students are packed closely together, one way to discourage cheating is to administer your test in multiple versions. Obviously, the version number needs to be included with the student answers. In order to score multiple test versions, you need to supply a correspondence file containing information on where each of the test items on an alternate version is located on the master version (which we will call version 1).

You may choose to supply the correspondence information in a text file or an Excel workbook. The example that follows has three versions of a five-item test, with the correspondence information placed in a text file, as shown here:

Correspondence Information in File c:\books\test scoring\corresp.txt

```
1 2 3 4 5
5 4 3 2 1
1 3 5 2 4
```

The first line of this file contains the item number on the first version. Line two contains the correspondence between versions 2 and 1. For example, the first item on the version 2 exam is item 5 on version 1, the second item on the version 2 exam is item 4 on version 1, and so forth. The third line contains correspondence information for version 3. The data in this text file is used to populate a two-dimensional temporary SAS array in this program.

The student data for this example is stored in a text file called mult_versions.txt. The first line of this file contains the answer key (for version 1). The remaining lines contain data for each student: a nine-digit ID, the test version (1, 2, or 3), and the student answers. A listing of this file is shown below:

Listing of Text File mult_versions.txt

```
          ABCDE
1234567891ABCDE
3333333331ABCDD
2222222222EDCBA
4444444441AAAAA
5555555553ACEBD
```

The program that follows uses the correspondence information to unscramble the student data so that it looks like every student took version 1 of the test:

Program 6.3: Scoring Multiple Test Versions

```
data multiple;
   retain Key1-Key5;
   array response[5] $ 1;
   array Ans[5] $ 1;
   array Key[5] $ 1;
   array Score[5];
   array correspondence[3,5] _temporary_; ❶
   if _n_ = 1 then do;
      *Load correspondence array;
      infile 'c:\books\test scoring\corresp.txt';
      do Version = 1 to 3; ❷
         do Item = 1 to 5;
            input correspondence[Version,Item] @;
         end;
      end;
      infile 'c:\books\test scoring\mult_versions.txt' pad; ❸
      input @11 (Key1-Key5)($upcase1.); ❹
   end;
   infile 'c:\books\test_scoring\mult_versions.txt' pad;
   input @1 ID $9. ❺
         @10 Version 1.
         @11 (Response1-Response5)($upcase1.);
   Raw = 0;
   do Item = 1 to 5;
      Ans[Item] = Response[correspondence[Version,Item]]; ❻
      Score[Item] =  (Ans[Item] eq Key[Item]);
      Raw + Score[Item];
   end;
   drop Item Response1-Response5;
run;
```

❶ The array CORRESPONDENCE has two dimensions (three rows and five columns) and is declared to be a temporary array by the key word _TEMPORARY_. The row and column dimensions are created by placing a comma between the 3 and 5 on the ARRAY statement. Temporary arrays are similar to real arrays in SAS. The only difference is that temporary arrays are not associated with any variables—only the array elements (that are stored in memory) are accessible in the DATA step.

The first dimension of the array (correspondence) is the version number; the second dimension of the array is the correspondence value. For example, suppose a student has version 2 of the test and she selects E as her answer to the first question. The program will select the first value in the second row of the correspondence table, which is a 5. The program then assigns the answer E to the variable Ans5.

❷ You load the CORRESPONDENCE array from the data in the corresp.txt file using nested DO loops. The outer loop cycles the version number and the inner loop cycles through the items.

❸ Once you have loaded the CORRESPONDENCE array, you then change the INFILE location to your test data. In this example, the test data is placed in a file called multi_versions.txt in a folder called c:\books\test scoring.

❹ You next read the key variables and student answers in the usual manner. The informat $UPCASE1. is used to read student answers instead of the usual $1. informat used previously. The $UPCASE informat, as you may have guessed, converts all character values to uppercase.

❺ You read in the student ID, the version number, and the student responses into variables that you name Response1 to Response5.

❻ This is the statement that places each student response, regardless of version, in the appropriate Ans variables. For example, for a student answering item 1 on a version 2 test, the value of `correspondence[2,1]` is `5`. Therefore, the variable Ans1 will be assigned the value of the variable Response5 for this student. In computer science, this method of looking up a value is called *indirect addressing*.

Here is the output from running this program:

Output from Program 6.3

Listing of MULIPLE

ID	Key1	Key2	Key3	Key4	Key5	Ans1	Ans2	Ans3	Ans4	Ans5	Score1	Score2	Score3	Score4	Score5	Version	Raw
123456789	A	B	C	D	E	A	B	C	D	E	1	1	1	1	1	1	5
333333333	A	B	C	D	E	A	B	C	D	D	1	1	1	1	0	1	4
222222222	A	B	C	D	E	A	B	C	D	E	1	1	1	1	1	2	5
444444444	A	B	C	D	E	A	A	A	A	A	1	0	0	0	0	1	1
555555555	A	B	C	D	E	A	E	D	C	B	1	0	0	0	0	3	1

To fully understand this, look at the original answers for student 222222222. This student took version 2 of the test and her answers were E D C B A—the values of the variables Ans1 to Ans5 are A B C D E.

Analyzing Tests with Multiple Versions (with Correspondence Information in an Excel File)

If you prefer to enter your correspondence information in an Excel file, you only need to make a small adjustment to Program 6.3. The format of your Excel spreadsheet needs to be structured as follows: The first row should contain column headings labeled Q1 to Q*n*, where *n* is the number of items on the test. The remaining rows should contain the same values as the text file in the previous section. A spreadsheet version of the previous text file is shown below:

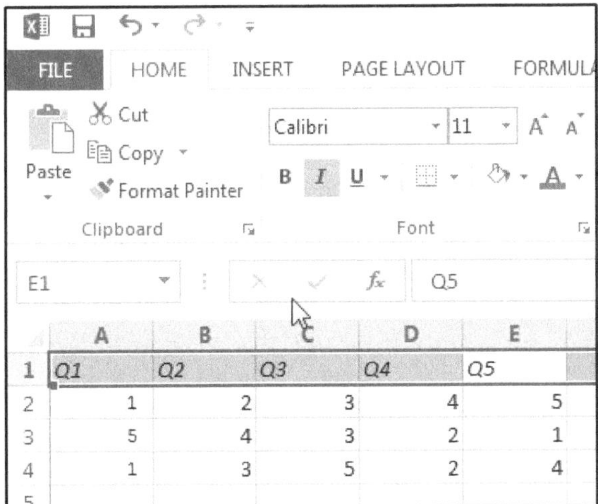

Although the bottom of this spreadsheet is not shown, the default name $Sheet1 was used.

A SAS program to read the correspondence information from the Excel file and score the test follows:

Program 6.4: Reading the Correspondence Information from an Excel File and Scoring the Test

```
libname readxl 'c:\books\test scoring\correspondence.xlsx';
data multiple;
   retain Key1-Key5;
   array Response[5] $ 1;
   array Ans[5] $ 1;
   array Key[5] $ 1;
   array Score[5];
   array Q[5];
   array Correspondence[3,5] _temporary_;
   if _n_ = 1 then do;
      *Load correspondence array;
      do Version = 1 to 3;
         set readxl.'Sheet1$'n;
        do Item = 1 to 5;
            Correspondence[Version,Item] = Q[Item];
         end;
      end;
      infile 'c:\books\test scoring\mult_versions.txt' pad;
      input @11 (Key1-Key5)($upcase1.);
   end;
```

```
   infile 'c:\books\test scoring\mult_versions.txt' pad;
   input @1 ID $9.
         @10 Version 1.
         @11 (Response1-Response5)($upcase1.);
   Raw = 0;
   do Item = 1 to 5;
      Ans[Item] = Response[correspondence[Version,Item]];
      Score[Item] =  (Ans[Item] eq Key[Item]);
      Raw + Score[Item];
   end;
   drop Item Response1-Response5;
run;
```

You read the Excel file using the LIBNAME engine (described in Chapter 2). Because you used column headings Q1 to Q5, you read the correspondence information from variables Q1 to Q5 in a DO loop. Also, since the default worksheet name $Sheet1 was used, you use that sheet name in the SET statement. If you choose another name for your worksheet, simply replace the name $Sheet1 with the name you have chosen. The resulting SAS data set is identical to the one in the previous section, so there is no need to show the listing again.

Analyzing Tests with Multiple Versions (with Correspondence Information and Student Data in an Excel File)

The last section of this chapter demonstrates how you can score multiple versions of a test when both your correspondence file and your student data are stored in Excel files. The correspondence file needs to be structured exactly as in the previous section. Your student data file needs to be structured as follows:

The first row of your spreadsheet should contain the column headings as follows:

Column A: ID

Column B: Version

Columns C – end: Student responses labeled R1 through R*n*, where *n* is the number of items on the test.

For example, your spreadsheet containing the same data as the text file described earlier is displayed next: (Leave the sheet name as the default value Sheet1.)

	A	B	C	D	E	F	G
1	ID	Version	R1	R2	R3	R4	R5
2			A	B	C	D	E
3	123456789	1	A	B	C	D	E
4	333333333	1	A	B	C	D	D
5	222222222	2	E	D	C	B	A
6	444444444	1	A	A	A	A	A
7	555555555	3	A	C	E	B	D

Here is a program to read data from these two Excel files and score the test:

Program 6.5: Reading the Correspondence Information and Student Data from Excel Files and Scoring the Test

```
libname readxl 'c:\books\test scoring\correspondence.xlsx';
libname readtest 'c:\books\test scoring\mult_versions.xlsx';

data multiple;
   retain Key1-Key5;
   array R[5] $ 1;
   array Ans[5] $ 1;
   array Key[5] $ 1;
   array Score[5];
   array Q[5];
   array Correspondence[3,5] _temporary_;
   if _n_ = 1 then do;
      *Load correspondence array;
      do Version = 1 to 3;
        set readxl.'Sheet1$'n;
        do Item = 1 to 5;
            Correspondence[Version,Item] = Q[Item];
        end;
      end;
      set readtest.'Sheet1$'n;
      do Item = 1 to 5;
         Key[Item] = R[Item];
      end;
   end;
   set readtest.'Sheet1$'n (firstobs=2);
   Raw = 0;
```

```
   do Item = 1 to 5;
      Ans[Item] = R[correspondence[Version,Item]];
      Score[Item] =  (Ans[Item] eq Key[Item]);
      Raw + Score[Item];
   end;
   drop Item R1-R5 Num_ID;
run;
```

Because the previous section describes how to use a LIBNAME engine to read Excel files, only a few brief comments are described. You write two LIBNAME statements, one for the correspondence file and one for the student data file. If you override the default sheet name Sheet1, you can substitute the name you chose for Sheet1 in the program. The variable names coming from the Excel file become the variable names in the SAS data set. Because the ID values are numeric in the Excel file, the variable ID in the SAS data set will also be numeric. If you wish, you can perform what SAS calls a *swap-and-drop operation,* where you rename the variable using a RENAME= data set option and then use a PUT function to create a character variable that you call ID.

The student data begins in column two, so you use the data set option FIRSTOBS=2 to read data values starting with the second observation. The resulting data set is identical to the two previous data sets with the exception that the ID variable will be numeric unless you decide to convert it.

Conclusion

Although you may feel that you should delete (and then rewrite) a multiple-choice item that has more than one correct answer, the first program in this chapter demonstrated how to allow for multiple correct answers.

At a medical school in New Jersey where one of the authors worked for 26 years, it was common practice to first run an item analysis on a test and then delete items that had poor item statistics (such as a difficulty that was too low or where the point-biserial coefficient was very low or negative). This practice often increases the overall test reliability. Program 6.2 demonstrated how to delete items and rescore the test.

The last topic of this chapter, scoring and analyzing multiple test versions, is of particular importance when you have a large number of students in an environment, such as a large lecture hall, where it is easy to copy answers from nearby students. Also, with cell phones and other communication devices, students are finding ways to cheat that we haven't even thought of yet. At the afore-mentioned medical school, we often administered three or four versions of mid-term and final exams. As you saw in Program 6.3, scoring and analyzing multiple test versions is easily accomplished. As a further method to discouraging cheating, Chapter 10 describes programs that can help you detect cheating on multiple-choice exams.

Chapter 7: Assessing Test Reliability

Introduction

Test reliability is a measure of repeatability. A test would not be very useful if, upon repeat administrations, a person's score varied widely. Intuitively, test reliability is related to test length. You would not expect a test of only a few items to be highly reliable because a person could make a few accurate guesses and obtain a high score or make a few bad guesses and obtain a low score.

One of the classic methods of assessing test reliability is called *test-retest reliability*. A group of subjects takes a test and, at some future time, typically a few weeks, retakes the same test. The time interval between the test and retest needs to be long enough so that the student does not remember her answer to each of the items but short enough so that the student has not either learned or forgotten material related to the test. You would expect the scores on the two administrations of the test to be related (correlated). The degree of correlation between the scores on the two test administrations is called test-retest reliability.

Because it is often impractical to administer a test a second time, several methods of assessing reliability have been developed that involve a single administration of the test. These methods include split-half reliability, the Kuder-Richardson Formula 20, and Chronbach's Alpha. These methods are discussed in this chapter, along with SAS programs to compute them.

Computing Split-Half Reliability

In the early days before computers were available, a common method of assessing test reliability was called *split-half reliability*. The concept was straightforward: You would split the test in half (usually by choosing the odd-numbered items as one test and the even-numbered items as a second test). You would then score each of the halves and compute the correlation between the two scores. Since longer tests tend to be more reliable than shorter tests, a formula developed by Spearman and Brown, called the *Spearman-Brown prophecy formula*, computes the reliability of a test that is shorter or longer than the one on which the reliability is measured. The formula is:

$$\rho^*_{xx'} = \frac{N\rho_{xx'}}{1+(N-1)\rho_{xx'}}$$

$$\rho^*_{xx'} = \frac{N\rho_{xx'}}{1+(N-1)\rho_{xx'}}$$

where $\rho^*_{xx'}$ is the estimated correlation for a test that is *N* times longer (or shorter) than the given test and $\rho_{xx'}$ is the correlation of the original test.

Because a split-half correlation is assessed on a test half the length of the original, you can apply the Spearman-Brown formula to make the adjustment, like this:

$$\rho^* = 2*\frac{\rho_{½}}{1+\rho_{½}}$$

where ρ^* is the estimated correlation and $\rho_{½}$ is the split-half correlation. Although this method is no longer used (now that we have computers), it is still interesting to compute this statistic. If nothing else, it makes for some interesting SAS programming.

The first step in computing a split-half correlation is to divide the test in half and score the two halves. Here is the program:

Program 7.1: Computing the Score for the Odd- and Even-Numbered Items

```
*Split-Half Reliability;
data split_half;
   set score;
   array ans[56] $ 1;
   array key[56] $ 1;
   array score[56];
   *Score odd items;
   Raw_Odd = 0;
   Raw_Even = 0;
   do Item = 1 to 55 by 2;
      Score[Item] = ans[Item] eq key[Item];
      Raw_odd + Score[Item];
   end;
   *Score even items;
   do Item = 2 to 56 by 2;
      Score[Item] = ans[Item] eq key[Item];
      Raw_Even + Score[Item];
   end;
   keep ID Raw:;
run;
```

A quick programming note: The expression Raw in the KEEP statement is a short-cut way of referring to all variables that begin with the characters Raw.

Here is a listing of the first five observations in data set SPLIT_HALF:

First Five Observations from Data Set SPLIT_HALF

Listing of data set SPLIT_HALF (first 5 observations)

ID	Raw_Odd	Raw_Even
203579875	25	22
116841443	21	20
176786926	22	23
011413555	19	20
051502147	21	20

You can now compute a correlation between the odd and even scores, using PROC CORR:

Program 7.2: Computing the Correlation Between the Odd and Even Scores

```
title "Computing the Correlation Between the Odd and Even Tests";
proc corr data=split_half nosimple;
   var Raw_Odd Raw_Even;
run;
```

Here is the result:

Computing the Correlation Between the Odd and Even Tests

The CORR Procedure

2 Variables:	Raw_Odd Raw_Even

Pearson Correlation Coefficients, N = 137 Prob > |r| under H0: Rho=0		
	Raw_Odd	Raw_Even
Raw_Odd	1.00000	0.69182 <.0001
Raw_Even	0.69182 <.0001	1.00000

The correlation between the Odd and Even items is .69182. The only task left is to adjust this correlation using the Spearman-Brown formula. This can be done by hand or in a short SAS DATA step. We will, of course, choose the programming solution, which looks like this:

Program 7.3: Computing the Spearman-Brown Adjusted Split-Half Correlation

```
*Using PROC CORR to output the split half value to
 a data set so that the Spearman-Brown prophcy formula
 can be applied;

proc corr data=split_half nosimple noprint
   outp=corrout(where=(_type_='CORR' and _NAME_='Raw_Odd'));
   var Raw_Odd Raw_Even;
run;

data _null_;
   file print;
   set corrout(keep=Raw_Even);
   Odd_Even_Adjusted= 2*Raw_Even / (1 + Raw_Even);
   put (Raw_Even Odd_Even_Adjusted)(= 7.3);
run;
```

You use the procedure option, OUTP=, to have PROC CORR output values to a SAS data set. This data set contains variables _TYPE_ and _NAME_, added by the procedure. If you wish to see the contents of this data set, you can run the PROC CORR portion of Program 7.3, without the data set options (that are used to subset the output data set to just those values you want). The correlation coefficient you want in the output data set has a value of _TYPE_ equal to 'CORR' and a value of _NAME_ equal to 'Raw_Odd'. To help understand this program, here is a listing of data set CORROUT, produced by PROC CORR:

Listing of Data Set CORROUT

Listing of data set CORROUT

TYPE	_NAME_	Raw_Odd	Raw_Even
CORR	Raw_Odd	1	0.69182

All that is left to do is to apply the Spearman-Brown formula in a DATA step and output the values of Raw_Even and the adjusted value of this coefficient.

An aside: Because you want to print out values in this DATA step but do not need the data set once you have printed the values, you use the special data set name _NULL_. This is a data set that is not a data set—that is, a SAS data set is not being created. Whenever you want to compute some values in a DATA step and plan to use PUT statements to output these values to files or to the output device, it is not necessary to create an actual data set, so using DATA _NULL_ is very efficient. By not creating a real data set, you reduce all the overhead required to write the data to a physical disk.

Here is the result of the PUT statement in the DATA _NULL_ step above:

```
Raw_Even=0.692 Odd_Even_Adjusted=0.818
```

The adjusted value for the split-half correlation is .818. That value is considered quite good for a typical in-class assessment test.

Computing Kuder-Richardson Formula 20 (KR-20)

Researchers G.F. Kuder and M.W. Richardson developed a formula to measure internal consistency reliability for measures with dichotomous choices. Known as the *Kuder-Richardson Formula 20*, it is similar to a split-half correlation, but it can be thought of as the mean split-half correlation if the test is divided in two in all possible ways. (This is not exactly true but makes for a good mental image of what is going on.) The KR-20 formula is:

$$r = \frac{K}{K-1}\left[1 - \frac{\sum_{i=1}^{K} p_i q_i}{\sigma_X^2}\right]$$

where K is the number of items on the test, p_i is the item difficulty (the proportion of the class that answered the item correctly), q_i is $1 - p_i$, and σ_X^2 is the variance of the test scores.

You can use PROC MEANS to compute the two variances used in the formula and output them to a SAS data set, as follows:

Program 7.4: Computing the Variances to Compute KR-20

```
proc means data=score noprint;
   output out=variance var=Raw_Var Item_Var1-Item_Var56;
   var Raw Score1-Score56;
run;
```

You use the keyword VAR= to output variances. Raw_Var is the variance of the test score (Raw), and the variables Item_Var1 to Item_Var56 are the item variances, which will be summed. You can now write a short DATA step to compute KR-20.

Program 7.5: Computing the Kuder-Richardson Formula 20

```
title "Computing the Kuder-Richardson Formula 20";
data _null_;
   file print;
   set variance;
   array Item_Var[56];
   do i = 1 to 56;
      Item_Variance + Item_Var[i];
   end;
      KR_20 = (56/55)*(1 - Item_Variance/Raw_Var);
      put (KR_20 Item_Variance Raw_Var) (= 7.3);
   drop i;
run;
```

You use a DO loop to sum the item variances. The expression Item_Variance + Item+Var[i] is a SUM statement because it is of the form VARIABLE + EXPRESSION. This has some differences from an ASSIGNMENT statement where you use an equal sign to assign a value to a variable. It is important to remember that a SUM statement has the following properties. First, the VARIABLE is automatically retained. Second, the VARIABLE is initialized to 0. Third, if the EXPRESSION is a missing value, it is ignored. The result of the DO loop is that the variable Item_Variance is the sum of the item variances. All that is left to do is to compute KR-20 and write it out to the Output window. The output looks like this:

```
Computing the Kuder-Richard Formula 20
KR_20=0.798 Item_Variance=7.902 Raw_Var=36.551
```

Computing Cronbach's Alpha

The KR-20 formula is applicable to tests where the item scores are dichotomous (right or wrong). A generalization of this formula, called *Cronbach's Alpha*, can be used for scaled items such as items on a Likert scale. Fortunately, you can use the ALPHA option with PROC CORR to compute this value. Cronbach's Alpha is equivalent to the KR-20 for dichotomous items. Here is the SAS code to compute Cronbach's Alpha, followed by the output:

Program 7.6: Computing Cronbach's Alpha

```
*Cronbach's Alpha;
proc corr data=score nosimple noprint alpha
   outp=Chronbach(where=(_type_='RAWALPHA') keep=_type_ Score1
   rename=(Score1=Alpha));
    var Score1-Score56;
run;
```

Output from Program 7.6

Listing of Chronbach

TYPE	Alpha
RAWALPHA	0.79805

Notice that the value of Chronbach's Alpha is identical to the KR-20 value computed earlier. Even if you have a test with dichotomous items, you may elect to use PROC CORR with the ALPHA option instead of the SAS code demonstrated in Programs 7.4 and 7.5.

Demonstrating the Effect of Item Discrimination on Test Reliability

There is a relationship between item discrimination and test reliability—the higher your item discrimination indices, the higher your overall test reliability will be. One way to demonstrate this is to rescore the statistics test we have been using, after removing four items with very low point-biserial coefficients. The following table shows the item statistics for these four items:

Four Items on the Statistics Test with Poor Item Statistics

Item Number	Difficulty	Point-Biserial Correlation	1st Quartile	2nd Quartile	3rd Quartile	4th Quartile
8	84%	.07	77.4%	82.9%	82.5%	93.3%
26	7%	.04	6.45%	5.71%	7.50%	6.45%
43	89%	.04	83.9%	91.2%	92.5%	87.1%
54	16%	-.02	19.4%	2.86%	27.5%	12.9%

Notice that the point-biserial correlation coefficients are either very close to 0 or negative. The original reliability for this test was as follows:

```
KR_20=0.798 Item_Variance=7.902 Raw_Var=36.551
```

You can use the macro programs in Chapter 11 to score the original test, rescore that test with the four items that performed poorly removed, and then compute the reliability of the rescored test. To start, you can use the SCORE_TEXT program to score the original statistics test as follows:

```
%score_text(%score_text(file='c:\books\test scoring\stat_test.txt',
        dsn=score_stat,
        length_id=9,
        Start=11,
        Nitems=56)
```

Next, use the %RESCORE macro to rescore this test as follows:

```
%Rescore(Dsn=Score_stat,
          Dsn_Rescore=New_Stat,
          Nitems=56,
          List=8 26 43 54)
```

Now that you have rescored the test, you can run the KR_20 macro to compute the Kuder-Richardson reliability, like this:

```
%KR_20(Dsn=New_Stat,
        Nitems=52)
```

Here are the results:

```
KR_20=0.818 Item_Variance=7.225 Raw_Var=36.503
```

The reliability, as measured by the KR-20, is higher, even though the test is now four items shorter.

Demonstrating the Effect of Test Length on Test Reliability

As discussed earlier, longer tests tend to be more reliable. To demonstrate this, the odd and even items of the 56-item statistics test were scored separately and the KR-20 was computed for each subtest, with the following results:

```
Computing the Kuder-Richardson Formula 20
Reliability of Test Selecting Odd Numbered Items
KR_20=0.685 Item_Variance=4.099 Raw_Var=12.063

Computing the Kuder-Richardson Formula 20
Reliability of Test Selecting Even Numbered Items
KR_20=0.626 Item_Variance=3.803 Raw_Var=9.599
```

The KR-20 for the entire test was .798. If we take the mean KR-20 from the odd and even test versions and apply the Spearman-Brown prophecy formula to estimate the reliability of a test twice as long, the result is .7915, quite close to the actual value.

Conclusion

Once you have completed your item analysis, it's time to estimate the overall test reliability. For dichotomous items (right or wrong), Kuder-Richardson's formula 20 and Cronbach's Alpha are equivalent and give you a good indication of your test's reliability. If you determine that either of these coefficients is too low, the next step is to attempt to rewrite items that have low point-biserial coefficients and, possibly, to increase the length of the test by writing new items.

Chapter 8: An Introduction to Item Response Theory - PROC IRT

Introduction

IRT stands for *Item Response Theory*, an approach to analyzing test data and building tests that has been in development since the 1960s. It is sometimes referred to as *modern test theory* as opposed to *classical test theory*, which employs concepts like Cronbach's coefficient alpha and item/test correlations. Although classical test theory is still widely used, especially in cases where the test is used only once and for situations where the number of examinees is small, IRT has become the dominant approach to test analysis in large testing programs (think SATs, MCATs, and international comparisons like PIRLS and TIMMS) and also in programs for certification of large numbers of students (the bar exam for lawyers, for example). Broadly speaking, if you have a test that is being given to less than 100 people, and for which you don't plan on generating an item bank to pull from, you are better off using classical test theory. If you are teaching a large lecture class and want to develop an item bank to make different versions of a test in different years, then you may want to explore IRT.

IRT starts with the notion that one is interested in measuring a single dimension, or trait, concerning individuals, such as quantitative ability or knowledge of a foreign language. IRT was often called *latent trait theory* when it was first introduced. Within IRT, the idea that you are measuring just one thing is known as the assumption of *unidimensionality*.

There are two major approaches to IRT, and they go by different names. There is what is known as the Rasch model, first presented by Danish mathematician Georg Rasch (1961) and popularized by Benjamin Wright (1977). The other approach generally is called IRT, and in this approach, the Rasch model is considered to be a special case of IRT. This approach was pioneered by Frederick Lord and Allan Birnbaum (see Lord, 1980; or Hambleton, Swaminathan, & Rogers, 1991, for an overview). A bit of a warning here: There are camps of scholars within each approach, and sometimes they can get rather critical of one another. The essence of the debate is between scholars who believe that measures should meet certain requirements (Rasch approach) versus scholars who believe in best capturing test data as they exist (general IRT approach).

The basic idea of IRT is that test items can be characterized by a limited number of parameters and that test items and examinees can be placed along the same ability/difficulty scale. Thus, in essence, a person can be smarter than an item is hard if his/her likelihood of getting that item correct is above a certain probability (which is often set at .5 but could be any probability). Furthermore, once test items have been analyzed and calibrated, any subset of items can be selected for administration to examinees, and an ability estimate can be derived that places examinees on the same scale. In fact, no two examinees have to take the same set of items once the items have been calibrated.

The difference between the Rasch model and the other IRT models has to do with the number of parameters used. The Rasch model only considers the item difficulty parameter; hence, it is sometimes referred to as the *one-parameter model*. This is abbreviated as 1PL, standing for one-parameter logistic model. Other IRT models employ two or three parameters (2PL, 3PL). The two-parameter model adds the strength of the relationship between the item and the underlying trait being measured as a second parameter (2PL) and an estimate of the influence of guessing the right answer as a third parameter (3PL). The 3PL model was developed specifically to model behavior on multiple-choice items, where guessing is a major consideration.

The fundamental idea of IRT can be understood by looking at the Rasch model. Each test item is characterized by how difficult it is to answer (item difficulty), and each person is characterized by how able he or she is (person ability). These characterizations are placed on the same scale. When the person's ability equals the item's difficulty, the person has a 50/50 chance of getting the item right. If the person's ability exceeds the item's difficulty, the person's chances of getting the item right increase in proportion to the difference between the person's ability and the item's difficulty. The Rasch model makes the testable assumption that all items bear roughly the same relationship to the overall trait and that guessing is not a strong factor in getting an item correct. When these assumptions are met, the Rasch model is an incredibly strong approach to test development.

The 2PL model does not assume that all items bear the same relationship to the overall trait; instead, the strength of that relationship is built into the measurement. And the 3PL model does the same with estimates of the influence of guessing.

It isn't possible to explore the intricacies of IRT here (or even come close!), but there are a number of excellent resources that explain the various approaches in detail. A good introduction can be found in DeMars (2010); Hambleton, Swaminathan, & Rogers, (1991); or Baker (2001), which has the substantial advantage of being available free online.

IRT basics

There are roughly four ideas behind IRT that you need to understand to get a rudimentary idea of what it's all about. These are assumptions and requirements of the models, item characteristics, estimation procedures, and fit statistics.

Assumptions and requirements of the models

IRT begins with an assumption that one is measuring a single trait or ability. This assumption can be tested by running a factor analysis on the test to see if the first factor is much greater than the second factor. This is roughly equivalent to saying that the coefficient alpha for the test is high. Both indicate that the items on the test appear to be measuring the same thing. Recent developments in IRT theory and practice have allowed for looking at data that is multidimensional, but that's beyond our discussion here. IRT also assumes that, except for the influence of a person's ability, responses to the test items are independent. In general, IRT requires larger sample sizes to obtain estimates of item parameters than one would find in classical test theory. For the Rasch model, one can often get good estimates with an *n* of 100, although larger samples are always better. For 2PL and 3PL, much larger sample sizes are usually needed.

Item characteristics

One way to think about the differences between IRT and CTT (classical test theory) is that IRT is much more about a set of items that serve as the basis for building tests for various purposes, whereas CTT is more about the characteristics of a given test where the items do not change. Thus, in CTT, we get an estimate of the reliability of the test, and we can calculate a standard error of estimate for a given administration of that test. In IRT, we have a bank of items, some of which we give to an individual in order to get an estimate of ability with a certain level of precision. Combined with computer administration, that estimate can be reached efficiently by choosing which item to give a person next based on the person's prior performance. We can set a level of precision in advance and continue the test session until that level is reached. Thus, what we really need in IRT is to know about the characteristics of the set of items we are using (at the item level), as opposed to what will happen with a given subset of those items (a given test).

Estimation procedures

There are a variety of ways to estimate the item characteristics required in IRT. They are all rather involved statistically, and we recommend you refer to the works listed above to learn about these approaches. It's not really necessary to master the nuances of those approaches to effectively use IRT.

Fit statistics

Whereas CTT uses reliability indices (such as coefficient alpha) to look at the quality of a test, IRT relies on fit statistics to see if the data fits the model (Rasch, 2PL, 3PL) that is being used and if individual items and examinees show good fit. If an item does not seem to fit the model, it is usually discarded. Lack of fit might be because there are aspects to the item that require abilities other than the trait under consideration. Sometimes, a person does not "fit" with the test experience. That is, a person might get a number of easier items wrong and harder items right. This could be due to indifference, language difficulties, cheating, or any of a host of possibilities. The point here is that the IRT approach allows for testing whether a person's encounter with a given set of test items seems to have produced a reasonable set of results.

SAS will produce fit statistics for the model as a whole and for individual items. This is a rather complex area, and we recommend that you consult one of the works mentioned earlier to get a strong understanding of how fit statistics work.

Looking at Some IRT Results

The most fundamental thing that we might look at in IRT is called an *item characteristic curve (ICC)*. It is a model of the relationship between how much ability a person has on the trait being measured and how likely they are to get a particular item correct. With the Rasch, or 1PL model, the only thing we need to know about an item in order to get a model of this curve is the item difficulty. The Rasch model requires that all curves have roughly the same slope and that guessing isn't a factor. The curve is a logistic ogive, and six of them are presented in Figure 8.1

You can see that the shape for these curves is identical but that they have been moved to the left or right, depending on how difficult the item was. The ability of persons is on the X axis on this graph, and the likelihood (probability) of getting the item right is on the Y axis. The scale on the X axis looks kind of like *z*-scores, but it could be transformed into any set of numbers. Item 4 (Score 4 in the figure) is the easiest of the six items. You can see that if a person's ability was at 1.0, the probability of getting the item correct is in the high .90s. Item 5 is the most difficult item. Here a person with an ability of 1.0 would only have about a .60 chance of getting the item correct.

Now, you might be asking, "Geez, how did the data get to be so smooth for these graphs? Do test items really look like that?" The short answer is "No." These aren't actual graphs of data, not even of smoothed data. They are curves of fixed characteristics that are laid in at the difficulty levels determined by the data. Now, since this is a set of graphs for dichotomously scored data (right = 1, wrong = 0), if we plotted all people in the data set individually, they would all have a Y value of 1 or 0. But if we grouped people into, say, six ability groups, we could get a better idea of whether the data for a given item really looks like the graphs presented below. If you think about it, for each of those observed data points, we would also have an expectation (given by the item characteristic curve), which would allow for a kind of chi-square analysis of fit. And that is what happens in IRT analysis, although again at a level of sophistication we can't get into here.

We might get one more useful idea out of the graphs of Figure 8.1. Imagine that we gave these six items to an examinee and that person got three of them correct. What would we estimate that person's ability to be? All we have to do is take an imaginary ruler, and move it from left to right on each of these items at the same time. If we stop that ruler at an estimate of -2.0, how many items would a person with that ability be expected to get right? To figure that out, we just have to go up the ruler to where it intersects with the curve and then read off the probability of getting that item right. For item 1, it looks like a person with an ability of -2.0 would have about a .55 chance of getting that item right. And then on item 2, the probability would be about .10, and then about .15 for item 3, etc. We could just add those probabilities together and find out, on average, how well we would expect a person with an ability of -2.0 to do on these six items together. To find out the ability of a person who got three items right, we just keep sliding the ruler to the right until the sum of the probabilities equals 3.0. It's that simple.

With the 2PL and 3PL models, the slopes of these curves can vary from one ICC to the next, and the lower asymptote (where the curve stretches out to the left) can be greater than 0 if guessing is involved.

What We Aren't Looking At!

IRT theory started with items that were scored dichotomously (right or wrong) and that only involved a single latent trait or ability. The theory has advanced dramatically over the years to be able to handle polytomous items (items scored on a scale) and multiple traits. These advances are both interesting and very useful but well beyond what we can address here.

Item Characteristics Curves from a One-Parameter Model

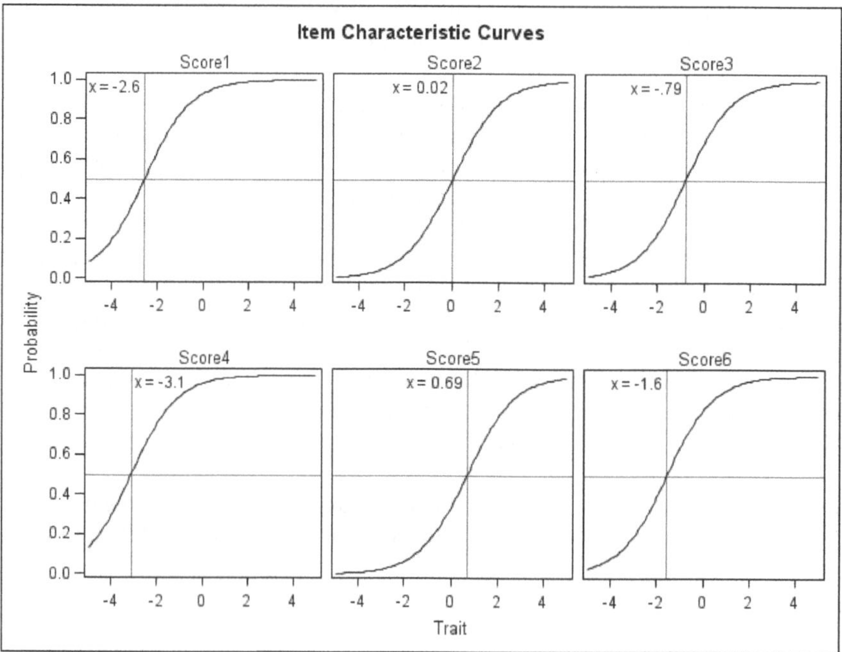

We now move on to how to use PROC IRT to analyze test data.

Preparing the Data Set for PROC IRT

In its simplest form, PROC IRT can model a series of dichotomous items with a unidimensional model. Program 8.1 below demonstrates how to run PROC IRT on the 56-item biostatistics test that was used to demonstrate classical item analysis in Chapter 5.

Program 8.1: Running PROC IRT for a Simple Binary Model

```
*Program 8.1;
*Preparing the stat test data set for PROC IRT;
data Binary;
   infile 'c:\books\test scoring\stat_test.txt' pad;
   array ans[56] $ 1 Ans1-Ans56;    ***student answers;
   array key[56] $ 1 Key1-Key56;    ***answer key;
   array score[56] Score1-Score56; ***score array 1=right,0=wrong;
   retain Key1-Key56;
   if _n_ = 1 then input @11 (Key1-Key56)($1.);
   input @11 (Ans1-Ans56)($1.);
   do Item=1 to 56;
      Score[Item] = Key[Item] eq Ans[Item];
   end;

   keep Score1-Score56;
run;

title "Listing of data set Binary (first 10 observations)";
proc print data=binary(obs=10);
run;
```

You start out by reading the answer key and the student answers as in previous programs. Here, however, there is no need to read the student IDs because PROC IRT is only looking for a set of binary (0/1) variables that represent, for each student, if an item was answered correctly (1) or incorrectly (0). You score the test in the usual way and keep the 56 Score variables.

Here are the first few observations in data set BINARY:

Listing of data set Binary (first 10 observations)

Obs	Score1	Score2	Score3	Score4	Score5	Score6	Score7	Score8	Score9	Score10	Score11	Score12	Score13	Sco
1	1	1	1	1	1	1	1	1	1	1	1	1	1	
2	1	1	1	0	1	1	1	1	0	0	1	1	0	
3	1	1	0	1	1	1	0	0	1	1	1	1	1	
4	1	1	1	0	1	1	0	1	1	1	0	0	0	
5	0	0	1	1	1	1	1	1	1	1	1	0	1	
6	1	1	1	1	1	1	1	1	1	1	1	1	1	
7	1	1	1	1	1	1	0	1	0	0	0	1	0	
8	0	1	1	1	1	1	1	1	1	1	0	0	1	
9	1	0	0	0	1	1	1	1	1	0	0	1	1	
10	0	1	1	0	1	1	1	1	1	1	1	0	0	

Running PROC IRT

The next step is to run PROC IRT as follows:

Program 8.2: Running PROC IRT (with All the Defaults)

```
*Program 8.2;
*Running PROC IRT (Unidimensional Model);
ods graphics on;
proc irt data=Binary;
   var Score1-Score56;
run;
```

When you run this program, you will see the following error message in the SAS Log:

```
*Program 8.2;
*Running PROC IRT (Unidimensional Model);
ods graphics on;

proc irt data=Binary;
   var Score1-Score56;
run;

ERROR: The number of levels for variable Score42 is smaller than 2.
NOTE: PROCEDURE IRT used (Total process time):
      real time            0.06 seconds
      cpu time             0.06 seconds
```

What went wrong? The error message is telling you that variable Score42 has fewer than two levels. This means that Score42 is all 0s (all students answered the item incorrectly) or all 1s (all students answered the item correctly – a more likely outcome). To be sure, you can run PROC FREQ as demonstrated in Program 8.3.

Program 8.3: Running PROC FREQ to Inspect Variable Score42

```
*Program 8.3;
*Identifying the Problem with Item 42;
title "Identifying the Problem with Item 42";
proc freq data=binary;
   tables Score42 / nocum nopercent;
run;
```

Output from PROC FREQ follows:

Identifying the Problem with Item 42

The FREQ Procedure

Score42	Frequency
1	137

As you can see, your guess that all the students answered item 42 correctly is confirmed. The statistical methods used to create test models (in this case, a unidimensional model) require that each variable must have two levels. You need to run PROC IRT with Score42 removed from the variable list, as shown next.

Program 8.4: Running PROC IRT with Score42 Omitted

```
*Program 8.4;
*Running PROC IRT with Item 42 Removed;
ods graphics on;
title "PROC IRT with Item 42 Removed";
proc irt data=Binary plots=(scree(unpack) icc);
   var Score1-Score41 Score43-Score56;
run;
```

In order to display some of the graphical output produced by PROC IRT, you need to do two things. First, you turn on ODS graphics (in versions of SAS 9.4 and higher, ODS graphics is turned on as the default and the ODS graphics statement is unnecessary). Second, you use a PLOTS= procedure option to request a scree plot (the UNPACK option places each of the two scree plots in a separate panel, rather than the two together) and an item characteristic curve (option ICC) for each item. Please note that even with relatively small tests taken by a few hundred students, this program may take considerable CPU time. For the sake of simplicity, we are shifting our analysis at this point to a 30-item physics test where we have a sample size of 102. Run Program 8.5, shown below, to run this model:

Program 8.5: Running a 1PL Model for a 30-Item Physics Test

```
title IRT Model on Physics Post Test;
proc irt data=test.all_post plots=(scree(unpack) icc);
   var Score1-Score30;
   model Score1-Score30;
run;
```

Here is the text portion of the output (broken down by sections):

IRT Model on Physics Post Test - All Defaults

The IRT Procedure

Modeling Information	
Data Set	TEST.ALL_POST
Link Function	Logit
Response Model	Graded Response Model
Number of Items	30
Number of Factors	1
Number of Observations Read	102
Number of Observations Used	102
Estimation Method	Marginal Maximum Likelihood

The modeling information presented here is simply a summary of what you've asked the program to do. Here we have 30 items and 102 people, and we're using maximum likelihood estimation. The next part of the output shows item information:

Item Information		
Item	Levels	Values
Score1	2	0 1
Score2	2	0 1
Score3	2	0 1
Score4	2	0 1
Score5	2	0 1
Score6	2	0 1
Score7	2	0 1
Score8	2	0 1
Score9	2	0 1
Score10	2	0 1
Score11	2	0 1
Score12	2	0 1
Score13	2	0 1
Score14	2	0 1

Because PROC IRT can be used for ordinal variables as well as binary ones, this section of output is included. With values of 0 or 1 for each item in this example, this portion of the output holds little interest.

IRT Model on Physics Post Test - All Defaults

The IRT Procedure

Eigenvalues of the Polychoric Correlation Matrix				
	Eigenvalue	Difference	Proportion	Cumulative
1	10.4598611	7.7082995	0.3487	0.3487
2	2.7515616	0.4044100	0.0917	0.4404
3	2.3471516	0.1124298	0.0782	0.5186
4	2.2347218	0.1321786	0.0745	0.5931
5	2.1025432	0.3430345	0.0701	0.6632
6	1.7595087	0.1677644	0.0587	0.7218
7	1.5917443	0.1423459	0.0531	0.7749
8	1.4493984	0.1901157	0.0483	0.8232
9	1.2592827	0.2175455	0.0420	0.8652
10	1.0417373	0.1138290	0.0347	0.8999
11	0.9279083	0.1021067	0.0309	0.9308
12	0.8258016	0.0824434	0.0275	0.9584
13	0.7433583	0.0474728	0.0248	0.9832
1[illegible]	0.6[illegible]	[illegible]	[illegible]	[illegible]

The first thing the program does is to run a factor analysis (principal components analysis) of the data to see if the underlying notion of a single trait being measured appears to be supported by the data. To check for this, we look at the eigenvalues. What you basically want to see here is the first eigenvalue being a lot larger than the second one. Then you want to see the third, fourth, etc., eigenvalues trailing off gradually after the second one. This indicates that you have a very strong first factor and that the remaining ones are pretty much indistinguishable. That appears to be the case in this situation. The first column is the eigenvalue itself, the second column is the difference between that eigenvalue and the next one (so for factor number one, we see that 10.4599 – 2.7516 = 7.7083). The third column tells you the proportion of the overall variance of the test items that is explained by that factor (10.4599/30 = 0.3487), and the last column gives you the cumulative proportion of variance that has been accounted for by that factor and the ones before it. Here we see a first eigenvalue that is four times larger than the second one, and the subsequent eigenvalues trail off gradually from the second one.

In some literature, you will see an argument for an eigenvalue greater than one criterion for how many factors there are in a data set. The logic there is that a factor should account for at least as much variance as an individual variable (or item) that comprises the factor analysis. But that criterion is way too liberal when a large number of variables (items) are in the factor analysis. A much better approach is called the *scree plot approach.* A scree plot is presented in Figure 8.2. It is a very simple plotting of the number of the eigenvalue (the first one, then the second one, etc.) on the X axis and the value of the eigenvalue on the Y axis. Thus, drawing from the table above, we see that the value in the scree plot for the first eigenvalue is 10.46, for the second one it's 2.75, and so on. A quick visual inspection of this plot shows that the first eigenvalue is substantially higher than any of the subsequent eigenvalues and that eigenvalues 2 through 30 just gradually reduce.

What we learn from looking at the scree plot is that the test items from the physics test under consideration have a strong first factor and that proceeding with the IRT analysis seems a good idea.

Scree Plot Showing the Eigenvalues for Each Factor

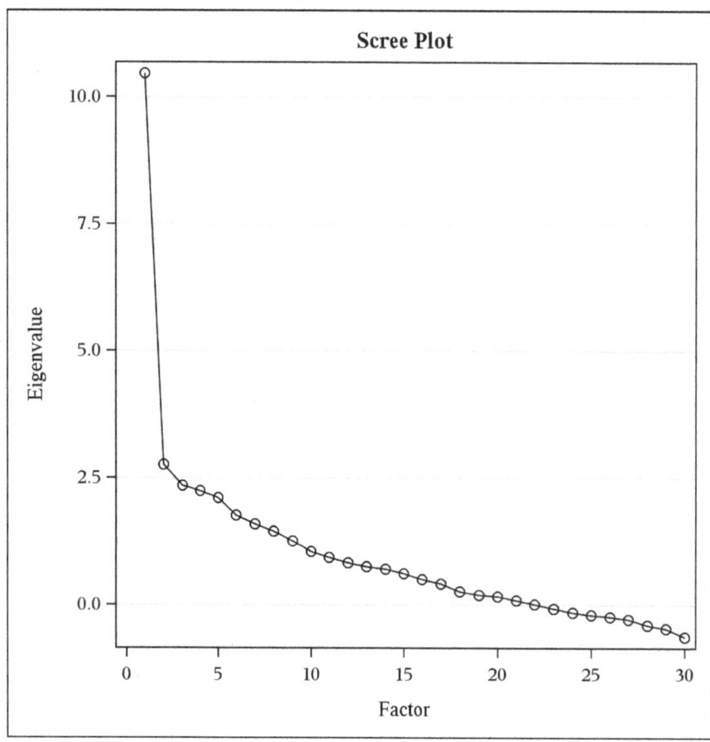

IRT Model on Physics Post Test - All Defaults

The IRT Procedure

Optimization Information	
Optimization Technique	Quasi-Newton
Likelihood Approximation	Adaptive Gauss-Hermite Quadrature
Number of Quadrature Points	19
Number of Free Parameters	60

The next part of the analysis shows details on how the parameters for the analysis were estimated. Quasi-Newton and adaptive Gauss-Hermite Quadrature approaches each have their own Wikipedia pages; knock yourself out.

IRT Model on Physics Post Test - All Defaults

The IRT Procedure

Item Parameter Estimates				
Item	Parameter	Estimate	Standard Error	Pr > |t|
Score1	Threshold	-2.23809	0.36467	<.0001
	Slope	0.58170	0.38576	0.0658
Score2	Threshold	0.03641	0.21576	0.4330
	Slope	0.62727	0.24844	0.0058
Score3	Threshold	-0.75969	0.26290	0.0019
	Slope	1.06535	0.32359	0.0005
Score4	Threshold	-2.88858	0.51629	<.0001
	Slope	0.84787	0.51278	0.0491
Score5	Threshold	0.88796	0.36800	0.0079
	Slope	2.21074	0.56898	<.0001
Score6	Threshold	-1.44637	0.30061	<.0001
	Slope	0.93179	0.34037	0.0031
Score7	Threshold	-0.88423	0.24641	0.0002
	Slope	0.75515	0.28542	0.0041
Score8	Threshold	1.44175	0.36857	

We now turn to the heart of the matter in looking at the output from the analysis. We have selected a 1PL model, which is basically the same idea as a Rasch analysis. The table above presents the item parameter estimates, which are essentially the critical pieces of information. In the table, we see columns for Item, Parameter, Estimate, Standard Error, and Pr > |t|. Item is the name we've given to the item, and Parameter tells us which parameter we are looking at for that item. (By "stacking" them, the program saves space horizontally.) Then we see the estimates, which are what we are after. For Score1, the threshold is -2.238 and the slope is 0.582. The numbering system for the threshold estimates are similar to *z*-scores. They run positive and negative and usually stay within +/-3.0. Although they are often a bit more spread out than *z*-scores, the *z*-score is not a bad analogy here. The *slope* is the slope of the relationship between ability and likelihood of a correct answer. The *threshold* is the point at which the probability of a correct answer would be .50 *if* the slope of the curve were equal to 1.0 (as it is assumed to be in the Rasch model). In the graphs shown below, the x value presented is the threshold divided by the slope, which provides the .50 level if the slope is assumed to be what the estimate says it is.

Some of the most interesting portions of the output are produced by ODS graphics. You see an item characteristic curve for each item on the test. If you look back at the output from Program 5.11 in Chapter 5, you see, at the far right, the proportion correct by quartile. If you plotted these values on the Y axis with the quartile values on the X axis, you would have a crude item characteristic curve. What PROC IRT does is model this curve from the test data and display it for you. Below are the item characteristic curves for the first six items. You need to keep in mind that these are idealized curves. The actual data will be much rougher in appearance than these curves.

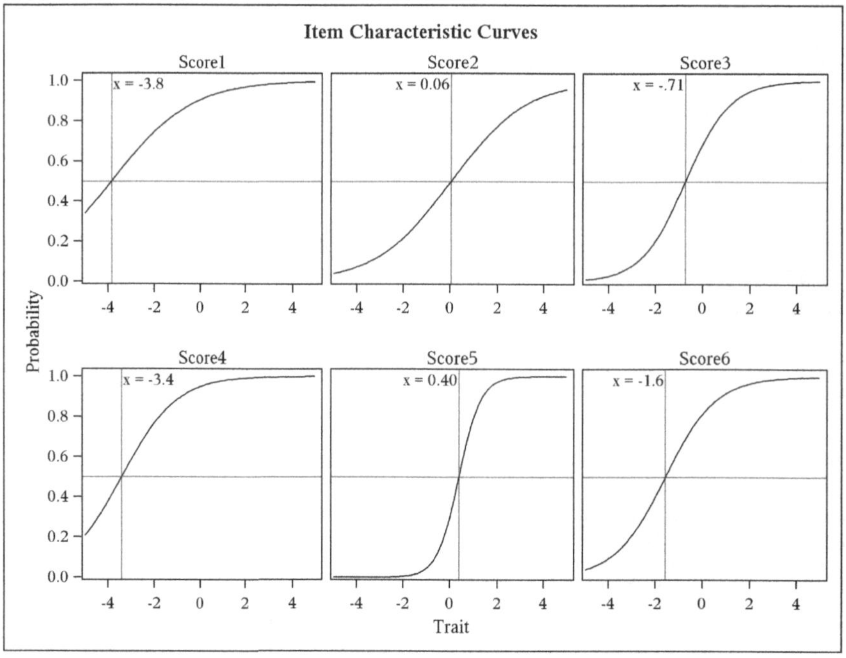

Running Other Models

To run other models, such as a Rasch, 2PL, or 3PL, you need to add a MODEL statement and specify the model you wish to run as an option. For example, to run a Rasch model with the 30-item physics test data, you proceed as follows:

Program 8.6: Running a Rasch Model on the 30-Item Physics Test

```
title IRT Model on Physics Post Test - Rasch Model;
proc irt data=test.all_post plots=(scree(unpack) icc);
   var Score1-Score30;
   model Score1-Score30 / resfunc=rasch;
run;
```

You use the RESFUNC= (response function) option to specify that you want to run a Rasch model. Other model options that can be specified with PROC IRT are:

ONEP - specifies the one-parameter model.

TWOP - specifies the two-parameter model.

THREEP - specifies the three-parameter model.

FOURP - specifies the four-parameter model.

GRADED - specifies the graded response model.

RASCH - specifies the Rasch model.

There are many other options, such as the rotation method for your factor analysis, that you can specify with PROC IRT. The SAS help facility provides a list of these options along with several examples.

Classical Item Analysis on the 30-Item Physics Test

Before we conclude this chapter, we thought you might be interested in seeing the results from running a classical item analysis on the physics test data. Macros from Chapter 11 were used to score the physics test and produce the item analysis output. A portion of the output showing the analysis on the first 15 items is shown next:

Item Statistics

	Choices							Quartile			
	A	B	C	D	E			1	2	3	4
	%	%	%	%	%	Diff.	Corr.	Prop. Correct	Prop. Correct	Prop. Correct	Prop. Correct
# Key											
1 C	4	2	90	4	.	90%	0.22	86.4%	87.0%	86.7%	100%
2 A	49	15	14	19	4	49%	0.33	22.7%	37.5%	70.0%	57.7%
3 C	15	16	65	1	4	65%	0.48	31.8%	54.2%	76.7%	88.5%
4 E	5	.	.	2	93	93%	0.24	77.3%	100%	96.7%	96.2%
5 B	17	36	14	25	9	36%	0.58	9.09%	20.8%	30.0%	80.8%
6 B	19	77	3	1	.	77%	0.38	59.1%	58.3%	90.0%	96.2%
7 B	11	69	6	4	11	69%	0.35	50.0%	45.8%	83.3%	88.5%
8 B	9	73	.	11	7	73%	0.52	28.6%	75.0%	83.3%	96.2%
9 E	1	19	22	7	52	52%	0.44	18.2%	45.8%	66.7%	69.2%
10 A	75	2	3	15	5	75%	0.58	36.4%	70.8%	86.7%	100%
11 D	7	6	28	58	1	58%	0.67	13.6%	37.5%	73.3%	96.2%
12 B	1	85	13	.	1	85%	0.33	59.1%	95.8%	83.3%	100%
13 D	3	6	34	57	.	57%	0.60	18.2%	20.8%	80.0%	96.2%
14 D	18	9	6	68	.	68%	0.48	40.9%	50.0%	73.3%	100%
15 A	67	7	26	.	.	67%	0.18	45.5%	79.2%	66.7%	73.1%

You see that item 1 was quite easy (90% of the students got it right) and item 2 was quite difficult (49% of the students got it right). This agrees with the threshold values in the IRT analysis, where the value for item 1 was -2.23809 and the threshold value for item 2 was.03641. That is, a low threshold indicates an easier item than a higher threshold. Take a look at the proportion correct by quartile and the point-biserial correlation for item 3. From a classical viewpoint, this is an excellent item. If you plotted the four proportion correct by quartile values (31.8%, 54.2%, 76.7%, and 88.5%), the resulting plot would have a similarity to the ICC curve produced by the IRT model. Seeing the IRT and classical output makes clear the two distinct purposes of these two methods of analysis—classical analysis is most useful for making decisions about individual items on a single test; IRT analysis is useful for calibrating items for item banks that will be used to assess student ability.

Conclusion

IRT is a fairly complicated approach to item analysis and test construction, and we have barely scratched the surface here. Our goal has been to provide an introduction to the analyses that SAS produces for IRT and to show how to run the procedure. To fully use IRT appropriately, you will need to consult the references listed (or similar ones) to develop a real sense of what IRT is about.

References

Baker, F. B. 2001. The basics of item response theory. ERIC Clearinghouse on Assessment and Evaluation. Retrieved 4/20/2014: ://echo.edres.org:8080/irt/baker/final.pdf

DeMars, C. 2010. *Item Response Theory*. Oxford: Oxford University Press.

Hambleton, R. K., Swaminathan, H., & Rogers, H. J. 1991. *Fundamentals of Item Response Theory*. Newbury Park, CA: Sage Press.

Lord, F. M. 1980. Applications of item response theory to practical testing problems. Mahwah, NJ: Lawrence Erlbaum Associates.

Rasch, G. 1961. On general laws and the meaning of measurement in psychology. In *Proceedings of the fourth Berkeley symposium on mathematical statistics and probability* (Vol. 4, pp. 321-333). Berkeley, CA: University of California Press.

Wright, B. D. 1977. Solving measurement problems with the Rasch model. *Journal of Educational Measurement*, 14(2), 97-116.

Chapter 9: Tips on Writing Multiple-Choice Items

Introduction

Testing has been conceptualized in a number of different fashions. It has been likened to taking a measure like height. It is related to a statistical sampling procedure where you have a conceptually infinite number of possible test items, and it has been described as trying to determine the shape of an object that resides within the brain. However it is conceptualized, a test consists of a set of one or more items. In the British tradition, testing often consists of a small number of essays (sometimes just one) on a topic. In America, and increasingly in the world, a test is thought of as a collection of individual items whose scores are summed together. Thus, testing requires the writing of test items. Although the primary focus of this book is how to analyze items and test scores once the test has been administered, we take some time here to talk about the construction of test items. This chapter is concerned with writing items for tests: course examinations, certification tests, admissions tests, or tests designed to assist in the learning process. What should a good test look like?

Tests, like all measures, should be fit for purpose. That is, they should do what you want them to do. Sometimes that will be an agglomeration of a bunch of different areas of your course, and other times it might be an attempt to create a scale that measures one well-defined ability, what measurement specialists often call a *trait*. There are many books that look at how to develop achievement tests and how to do measurement theory in general. We would recommend Haladyna and Rodriguez (2013) for test development; for measurement theory in general, see Brennan (2006).

Getting Started/Organized

Before you start writing items themselves, it is often useful to step back and get organized for the test.

Test Blueprints

It is very helpful to start an achievement measure with some sort of outline of what you want to accomplish. This is sometimes referred to as a *test blueprint*. A blueprint should be specific enough that if given to a colleague in the same field, that person could do a good job of writing the test. It does not need to be more explicit than that. Not only is the test blueprint very helpful for writing the test, it allows you to take a look at the blueprint and determine whether you are really testing students on the material that you feel is important. You should be able to look at a blueprint and conclude, "Yep, that's the course, the whole course, and nothing but the course."

A blueprint is simply an elaborated outline of what will be on the test. There are approaches that use a kind of matrix design for this activity, but that isn't necessary. Once the outline has been constructed, you can look at it and assign weightings to the various components of the test. We always have our weightings sum to 100%, but that isn't an absolute, depending on how you construct your overall grading system.

It is also often very useful to distribute the test blueprint to students in the course in advance of the test. That way you know that their studying will be focused on those aspects of the test that are important to you. It also eliminates the "you tested us on things you didn't teach us" and "you taught us stuff that wasn't on the test" complaints. Some faculty get concerned that one is spoon-feeding students by providing this information. We look at it from the opposite side of the coin: Once you have told students what to study, you are pretty much free to write a rigorous assessment of that material. It allows you to write a test that, if students do well on it, you will be pleased to assign them good grades.

Taxonomy of Objectives and Items

A second useful starting point is to think about the level of difficulty and complexity you want your items to have. What do you want your students to do? Should they be able to recall facts and figures, should they comprehend the material presented (restate it in their own words, perhaps?), should they be able to contrast a given theory to alternatives? One way to think rigorously about issues such as this is to consult a taxonomy of objectives and/or test items. The idea of a taxonomy of educational objectives (what you want to accomplish in a course) was first proposed by Bloom (1956), and a useful revision of that has been developed by his colleagues, Anderson and Krathwohl (2001). The revised taxonomy has six levels of complexity, or levels of thinking, into which course objectives and test items might be classified (the original taxonomy also had six levels). Very briefly, these levels are:

- **Remembering:** Simple recall of information such as facts and figures.
- **Understanding:** Does the student understand the information? Can he/she rewrite it in his/her own terms?

- **Applying:** Can the student take the information or ideas presented and use them in a novel setting? Can the student apply the information to a new task?
- **Analyzing:** Can the student take the ideas/theory/information apart and analyze components, compare to other ideas?
- **Evaluating:** Can the student make a critical judgment about the theory/ideas and defend that judgment?
- **Creating:** Can the student create a new idea/product/concept that is applicable to the task or situation at hand?

There are a variety of web sites that present information on this taxonomy and others that have been developed since Bloom's ground-breaking work. The underlying idea here is to think about levels of understanding and ability that go beyond remembering and understanding information. We are particularly enamored of the application level of the taxonomy. Can the student take what has been learned and use it in a new situation?

There are probably other things that one should do before writing a test, but this is a good beginning for now, so we will move on to actually looking at different possibilities for test items and how to write them.

Types of Items for Achievement Tests

Very broadly speaking, one can consider two basic types of test questions: ones that require recognition of correct answers and ones that require the generation of correct answers. There are a variety of formats possible within these two broad categories, and we will examine a number of them.

Recognition format simply means that the correct answer is presented to the examinee, who has to identify it among distracters, determine whether it is true or false, or match one characteristic to another (body parts and their names, for example). This format has the distinct advantage of being very easy to score, both in terms of assigning a value to a response (correct or incorrect) and the ability to use machine-based scoring of responses. It has several disadvantages, the main one being that examinees can guess the correct answer. Another important disadvantage is that it is somewhat limiting in terms of the type of abilities that can be elicited (although we will see that this limitation is not as severe as some believe).

Multiple-Choice Items

The multiple-choice (MC) item is ubiquitous in American education and becoming more popular worldwide. The MC item has basically two components, the stem, or question, which poses the problem; and the options, or distracters, which are presented as a set of typically three to five choices. The examinee reads the stem and has to select the option that best answers the question posed in the stem. Some examples of MC stems:

- **When did World War I begin?** This stem requires recalling information, the lowest level of the taxonomy. The student simply has to remember when the war began.
- **How did the killing of Archduke Ferdinand provide the spark for the war to begin?** This stem requires the student to understand why the assassination of Archduke Ferdinand caused the war to begin.
- **Which of the following "hotspots" in the world today might be thrown into war with the assassination of a world leader?** This question requires the student to understand the situation in a number of places in the world and apply what he/she knows about the causes of armed conflict to those situations.

In the examples provided here, it is not difficult to see that the three stems listed here not only vary in their level of the taxonomy but also in their difficulty. That is, the higher the level of the taxonomy, the more difficult the question. One often, but not always, sees that relationship.

But we don't yet really know the difficulty of these items. Indeed, each of them could be posed in generative, or what is often called *constructed response*, format. That is, they could just be presented as is, and the student would have to generate an answer for them. That would make them much more difficult than if presented in multiple-choice format. But, in multiple-choice format, the choice of options to use greatly affects the difficulty of the item. Consider the first stem above:

When did World War I begin?

Now, one could write a fairly easy set of options:

a) *1914*
b) *1941*
c) *1956*
d) *1968*

Or, one could write an incredibly challenging set of options:

a) *July, 1914*
b) *August, 1914*
c) *September, 1914*
d) *October, 1914*

Although most people would feel that the second set of choices might be a bit unfair, the point here has to do with what you expect the students to know. If you want them to generally understand that the war started roughly in 1914, you might use the following options:

a) *1912*
b) *1914*
c) *1915*
d) *1917*

That gives a reasonable set and will measure whether a student understands that the war started in 1914. Well, it started in 1914 in Europe. The US didn't join until 1917, and so 1917 may be considered by some students to be an unfair choice! One of the rules of writing MC items is to make sure that the right answer is right and that the wrong answers are wrong!

Another format for the MC item that is very useful is to start with a setting or context or a set of information that sets up the question. This can be done on a question-by-question basis, or one can set up a more elaborate setting and ask several questions about it (think of the reading comprehension questions from the SAT where a passage is presented and then four or more questions are asked about it). The use of introductory material in this fashion can often allow the test constructor to ask questions requiring the use of higher order thinking skills – those at the top of the taxonomy presented before. Imagine that we wanted to measure how well you have understood some of the more sophisticated programming issues presented in this text. We could present an example of some code used to solve a particular analysis problem. And that code could have an error in it. We could ask you to analyze the code and locate the error. Depending on how subtle we made the error, this could be a very difficult problem, requiring a high level of knowledge and analytical skill. Here is an example:

Given the following SAS program to read multiple observations from one line of data, identify the line that contains an error:

```
a)  data temperature;
b)     input Temp_C;
c)  Temp_F = (9*5)*Temp_C + 32;
d)  datalines;
e)  15 16 17 18 19 20
      run;
```

The answer is b. You need a double-trailing at sign (@) before the semi-colon to prevent the program from moving to a new line of data at each iteration of the DATA step.

Another possibility is to display a diagram or illustration, and then pose questions about it. However, when writing multiple questions about a single source or context, make sure that you are not giving away the answer to a question by the way another question is posed.

Giving away answers to questions brings up the issue of general do's and don'ts for writing MC items. One can find a number of these lists (some quite long) on the Internet, but here is a shortened version that focuses on key elements of good item writing:

- **Alignment:** Make sure each item is important and can be clearly linked to your test blueprint. Ask yourself, "Do I really want to know if the examinee can do this?"
- **Item Baggage:** *Item baggage* is the ease with which a person who knows what you are interested in can get the item wrong. Avoid double negatives – actually, it is best to avoid all negatives; they introduce noise into the answering of the item. Avoid stems like, "Which of the following is not used in… ." The general rule is no tricks. If an examinee gets an item wrong, you want it to be for one reason and one reason only: the person did not know the material.
- **Clear Problems:** Put the information necessary to do the problem/question in the stem of the item. That is, try to avoid overly long choices in the option list. You want the examinee to understand the nature of the problem once he/she has read the stem.
- **Clearly Correct:** As mentioned above, make certain that your right answers are right and your wrong answers are wrong. The easiest way to do this is to give the test to a colleague to answer. You will be amazed at how often something you thought was perfectly clear confuses someone who knows as much about the material as you do!
- **Parsimony:** Don't repeat the same phrase time and again in the option list. Just put it up in the stem.
- **Similarity of Options:** The similarity of the choices in a multiple-choice item often determines how difficult the item is: the more similar, the more difficult. Make sure the surface features in your choices are roughly similar. Don't have some choices use extreme language ("always," "never," "could not possibly," etc.) and others use very moderate language ("might," "can sometimes be seen to," etc.). Consistency is the key.
- **Grammar:** Make sure all choices are grammatically consistent with the stem.
- **Number of Choices:** Don't force yourself to have five MC items or even four. For course exams, you also don't need the number of choices to be the same for all items.

Examples of Strong and Poor MC Items

MC items are the workhorse of the examination field. Let's take a look at several and make some commentary on their quality.

Which of the following best describes the transmission of information from one neuron to another:

A. The axon of one neuron is attached to the dendrites of a number of postsynaptic neurons, which lets electrical current be passed among neurons.
B. An electrical impulse down the axon of one neuron causes neurotransmitters to be sent across a synaptic gap and received by dendrites of another neuron.
C. The nucleus of the first cell divides, and the information contained in the DNA of that cell passes on to the DNA of the second cell through the synaptic cleft.

D. The axon of one neuron sheds its outermost glial cell, which is received by the postsynaptic neuron through the transmitters contained at the end of the dendrites of the receiving cell.

This item, although it seems to be pretty substantial, suffers from three problems. First, the nature of the task really resides as much in the alternatives as it does in the problem itself. The examinee has to read all the choices and make comparisons in order to answer the question. This isn't always a bad thing, but it is when related to the second problem this item has: the choices are very long and complex. Making the comparisons among complex alternatives is not what is supposed to be being measured here. Instead, we want to know about the examinee's understanding of how neurons work. Finally, the incorrect choices aren't really reasonable (plausible) answers to the question.

A second example:

Three-year-old Maria's parents are native speakers of two different languages. They each speak to Maria in their native languages. Maria is learning to speak both languages fluently. What is this an example of:

A. Experience-expectant learning
B. Tabula rasa
C. Functional organization
D. Grammatical processing

This second item is an example of applying a concept. The examinee has to be able to interpret a new situation and determine which of the four concepts listed it describes. The problem is clearly stated in the stem, and the choices are plausible, while clearly being incorrect.

And a third example:

The pegword method is a good approach for learning lists of items. What is the essential feature of the pegword method that makes it an effective mnemonic device:

A. Rehearsal
B. Modeling
C. Disinhibitory effects
D. Visual imagery

This third example requires the examinee to look at the characteristics (features) of the pegword method and determine which of those characteristics makes it work as a mnemonic device. Exactly what level of difficulty this item exists at depends a bit on what has been taught in the course. If the features and why they work have been taught explicitly, then this is a recall type item. If the pegword approach has been taught, but not explained, then this item is at least at an understanding level. If the basic ideas behind mnemonic strategies and the information processing theory that lies behind it have not been taught, one could argue that this is an analyzing item. Again, the level of thinking involved depends on what has gone on in the course.

Conclusion

The best place to start in analyzing test data is with a good test! Test analysis will help to point out a number of areas where there are weaknesses with a test, or a test question, or even just an aspect of a test question. But it is essential to start with your very best effort. In this brief chapter, we look at the most commonly used examination question, the multiple-choice question, and provide some help on how to organize yourself for the test and how to think about writing test items, as well as some practical help on test construction.

References

Anderson, L. W., and Krathwohl, D. R. Eds. 2001. A Taxonomy for Learning, Teaching and Assessing: A Revision of Bloom's Taxonomy of Educational Objectives: Abridged edition, New York: Longman.

Bloom, B. S. 1956. Taxonomy of educational objectives. Vol. 1: *Cognitive domain.* New York: McKay.

Brennan, R. L. Ed. 2006. *Educational Measurement, 4th ed.* ACE/Praeger Series on Higher Education. MD: Rowman and Littlefield.

Haladyna, T. M., & Rodriguez, M. C. 2013. *Developing and Validating Test Items.* New York, NY: Routledge.

Chapter 10: Detecting Cheating on Multiple-Choice Tests

Introduction

This chapter covers several methods on how to detect cheating on multiple-choice exams. It is important to investigate claims of cheating very carefully since an accusation of cheating has such serious consequences. We recommend that the methods and programs presented here be used only after a question of cheating has been raised by independent means, such as suspicious behavior observed by a proctor or other independent evidence that cheating occurred. This is important because if you look at every student in a large class, the probability of finding a possible cheating incident increases (analogous to an experiment type I error in hypothesis testing).

The methods presented here look at the pattern of wrong answer choices. The first method discussed looks at the set of wrong answers from one student (whom we will call Student 1) and then counts the number of students who chose the same wrong answers to this set of items. This method, and a similar method that uses a set of items that two students got wrong in common (called *joint-wrongs*), requires that there be enough difficult questions on the test. If a test is too easy (or too short), the set of wrong answers or joint-wrongs will not be large enough for a meaningful analysis. The normal *p*-values used in statistical tests (alpha = .05 or .01) are not used in this type of investigation. Typically, *p*-values of less than 10^{-5} or lower are used.

How to Detect Cheating: Method One

This first method, described in a paper (Cody, 1985), uses a set of wrong answers by Student 1 and counts the number of the same wrong answers with every member of the class (with the exception of Student 2, who was possibly involved in cheating). The mean and standard deviation are computed. The distribution of the number of the same wrong answers is approximately normal (this can be tested), so you can compute a *z*-score and *p*-value for the number of the same wrong answers between Students 1 and 2. The probability of two students selecting the same wrong answers also depends on each student's ability. Two weak students are more likely to answer an item incorrectly and are, therefore, more likely to select the same wrong answer than two strong students. As a check, you can run the analysis on all students with grades close to Student 2 and inspect the *z*-score and *p*-value for these students.

Before you begin examining the SAS program that implements this method of detection, let's focus on a very simple data set (test_cheat.txt) to see exactly how the analysis works. Here is a listing of test_cheat.txt:

Listing of Data Set test_cheat.txt

```
   ABCDEA
001AAAAAA
002ABCEEA
003AAAEEA
004ABCAAC
005AAAAAA
```

The table below shows the number of wrong answers for each student as well as the number of items where a student chose the same wrong answer as Student 1:

Table Showing Wrong Answers and Same Wrong Answers for Each Student

Key or ID	Item 1	Item 2	Item 3	Item 4	Item 5	Item 6	# of Same Wrong
Key	A	B	C	D	E	A	
001	**A**	**|A|**	**|A|**	**|A|**	**|A|**	**A**	
002	A	B	C	|E|	E	A	0
003	A	||A||	||A||	|E|	E	A	2
004	A	B	C	||A||	||A||	|C|	2
005	A	||A||	||A||	||A||	||A||	A	4

Table Key: Single bars |X| indicate a wrong answer; double bars ||X|| indicate the same wrong answer as 001

In this example, Student 001 is Student 1, with four wrong answers. The last column shows the number of items where the student chose the same wrong answer as Student 1. The programs described in this chapter do not require that the data for Student 1 be the first record in the file (following the answer key).

Only a general discussion of the program next is provided.

Program 10.1: Program to Detect Cheating: Method One

```
*Macro Compare_Wrong uses the set of wrong answers from ID1
 and computes the number of same wrong answers for all students
 in the data file.  It then computes the mean and standard
 deviation of the number of same wrong answers (variable Num_Wrong)
 with both ID1 and ID2 removed from the calculation.  Finally,
 it computes a  z-value for the number of same wrongs for student ID2;

%macro Compare_wrong ❶
  (File=,           /*Name of text file containing key and test data */
   Length_ID=,      /*Number of bytes in the ID            */
   Start=,          /*Starting column of student answers */
   ID1=,   /*ID of first student */
   ID2=,   /*ID of second student */
   Nitems= /*Number of items on the test */ );

    data ID_one(keep=ID Num_wrong_One Ans_One1-Ans_One&Nitems ❷
        Wrong_One1-Wrong_One&Nitems)
        Not_One(keep=ID Num_wrong Ans1-Ans&Nitems
        Wrong1-Wrong&Nitems);

   /* Data set ID_One contains values for Student 1
      Data set Not_one contains data on other students
      Arrays with "one" in the variable names are data from
      ID1.
   */
      infile "&File" end=last pad;
   /*First record is the answer key*/
      array Ans[&Nitems] $ 1;
      array Ans_One[&Nitems] $ 1;
      array Key[&Nitems] $ 1;
      array Wrong[&Nitems];
      array Wrong_One[&Nitems];
      retain Key1-Key&Nitems;
      if _n_ = 1 then input @&Start (Key1-Key&Nitems)($1.); ❸
      input @1 ID $&Length_ID..
            @&Start (Ans1-Ans&Nitems)($1.);
      if ID = "&ID1" then do; ❹
         do i = 1 to &Nitems;
            Wrong_One[i] = Key[i] ne Ans[i];
            Ans_One[i] = Ans[i];
         end;
         Num_Wrong_One = sum(of Wrong_One1-Wrong_One&Nitems);
         output ID_one;
         return;
      end;
```

```
      do i = 1 to &Nitems; ❺
         Wrong[i] = Key[i] ne Ans[i];
      end;
      output Not_One;
   run;

   /*
   DATA step COMPARE counts the number of same wrong answers as
Student
   ID1.
   */
   data compare; ❻
      if _n_ = 1 then set ID_One;
      set Not_One;
      array Ans[&Nitems] $ 1;
      array Ans_One[&Nitems] $ 1;
      array Wrong[&Nitems];
      array Wrong_One[&Nitems];
      Num_Match = 0;
      do i = 1 to &Nitems;
         if Wrong_One[i] = 1 then Num_Match + Ans[i] eq Ans_One[i];
      end;
      keep Num_Match ID;
   run;

   proc sgplot data=compare; ❼
      title 'Distribution of the number of matches between';
      title2 "Students &ID1, &ID2, and the rest of the class";
      title3 "Data file is &File";
      histogram Num_Match;
   run;

   /*
   Compute the mean and standard deviation on the number of same
   wrong answers as ID1 but eliminate both ID1 and ID2 from the
   calculation
   */
   proc means data=compare(where=(ID not in ("&ID1" "&ID2")))noprint;
❽
      var Num_Match;
      output out=Mean_SD mean=Mean_Num_match std=SD_Num_match;
   run;
```

```
   data _null_; ❾
      file print;
      title1 "Exam file name: &File";
      title2 "Number of Items: &Nitems";
      title3 "Statistics for students &ID1 and &ID2";
      set compare (where=(ID = "&ID2"));
      set mean_sd;
      set ID_One;

      Diff = Num_Match - Mean_Num_Match;
      z = Diff / SD_Num_match;
      Prob = 1 - probnorm(z);

            put // "Student &ID1 got " Num_wrong_One "items wrong" /
             "Students &ID1 and &ID2 have " Num_Match "wrong answers
in common" /
             "The mean number of matches is" Mean_Num_Match 6.3/
             "The standard deviation is" SD_Num_match  6.3/
             "The z-score is " z 6.3 " with a probability of" Prob;
   run;
      file print;
      title1 "Exam file name: &File Number of Items: &Nitems";
      title2 "Statistics for students &ID1 and &ID2";
      set compare (where=(ID = "&ID2"));
      set mean_sd;
      set ID_One;

      Diff = Num_Match - Mean_Num_Match;
      z = Diff / SD_Num_match;
      Prob = 1 - probnorm(z);

      put // "Student &ID1 got " Num_wrong_One "items wrong" /
      "Students &ID1 and &ID2 have " Num_Match "wrong answers in
common" /
      "The mean number of matchs is " Mean_Num_Match /
      "The standard deviation is " SD_Num_match  /
      "This gives a z-score of " z " and a probability of " Prob;
   run;

   proc datasets library=work; ❿
      delete ID_One;
      delete Not_One Means;
   quit;

%mend compare_wrong;
```

❶ The calling arguments to this macro are the name of the text file containing the test data, the length of the student ID field, the position in the file where the student answers start, the two IDs of interest, and the number of items on the test.

❷ Two data sets are created. ID_One contains the data for ID1. Variables in this data set include the answers to each of the items (Ans_One variables), which items were answered incorrectly (Wrong_One variables equal to 1), and the number of wrong answers by ID1 (Num_Wrong_One). Data set Not_One contains the test data for all of the other students in the class.

❸ You read the answer key into the key variables and then read in student data.

❹ If the ID is equal to ID1, store the answers in the Ans_One variables and set the Wrong_One variables to 1 for all wrong answers (and 0 for correct answers). Then output these values to the ID_One data set.

❺ For all the other students, score the test in the usual way.

❻ Compute the number of the same wrong answers for each of the students.

❼ Use PROC SGPLOT to produce a histogram of the number of the same wrong answers.

❽ Compute the mean and standard deviation for the number of the same wrong answers, excluding Students 1 and 2.

❾ Compute the *z*- and *p*-values and produce a report.

❿ Delete work data sets created by the macro.

Let's first test the macro using the test_cheat.txt file described previously. For this file, the length of the ID field is 3, and the student answers start in column 4. If Student 001 is to be ID1 and Student 005 is to be ID2, the calling sequence looks like this:

Calling Sequence for Macro %Compare_Wrong

```
%Compare_Wrong(File=c:\books\test scoring\test_cheat.txt,
                 Length_ID=3,
                 Start=4,
                 ID1=001,
                 ID2=005,
                 Nitems=6)
```

Here are the results:

Output from Macro %Compare_Wrong

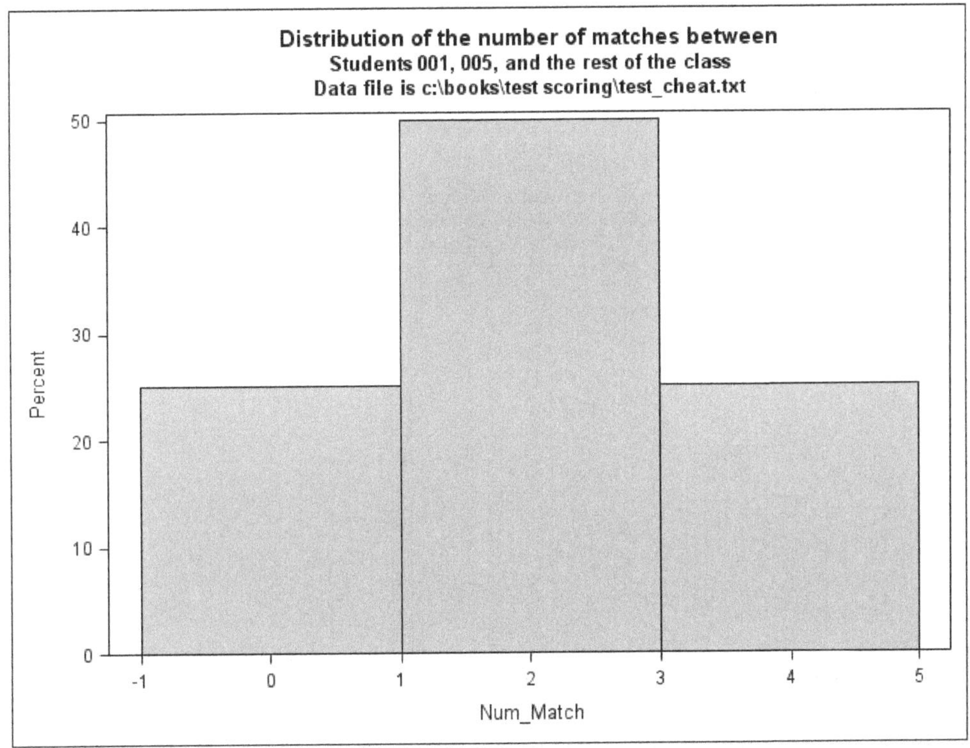


```
Exam file name: c:\books\test scoring\test_cheat.txt
Number of Items: 6
Statistics for students 001 and 005


Student 001 got 4 items wrong
Students 001 and 005 have 4 wrong answers in common
The mean number of matches is 1.333
The standard deviation is 1.155
The z-score is  2.309 with a probability of
0.0104606677
```

You can verify that the number of matches, the mean, and the standard deviation of the number of matches are consistent with the sample data in the table presented earlier in this chapter.

For a more realistic test, the test data from the 56-item statistics test was doctored so that two students had similar wrong answers. The two students with modified data were ID 123456789 and ID 987654321. Calling the %Compare_Wrong macro with this information looks like this:

Calling the %Compare_Wrong Macro with IDs 123456789 and 987654321 Identified

```
%compare_wrong(File=c:\books\test scoring\stat_cheat.txt,
               Length_ID=9,
               Start=11,
               ID1=123456789,
               ID2=987654321,
               Nitems=56)
```

Output from the %Compare_Wrong Macro

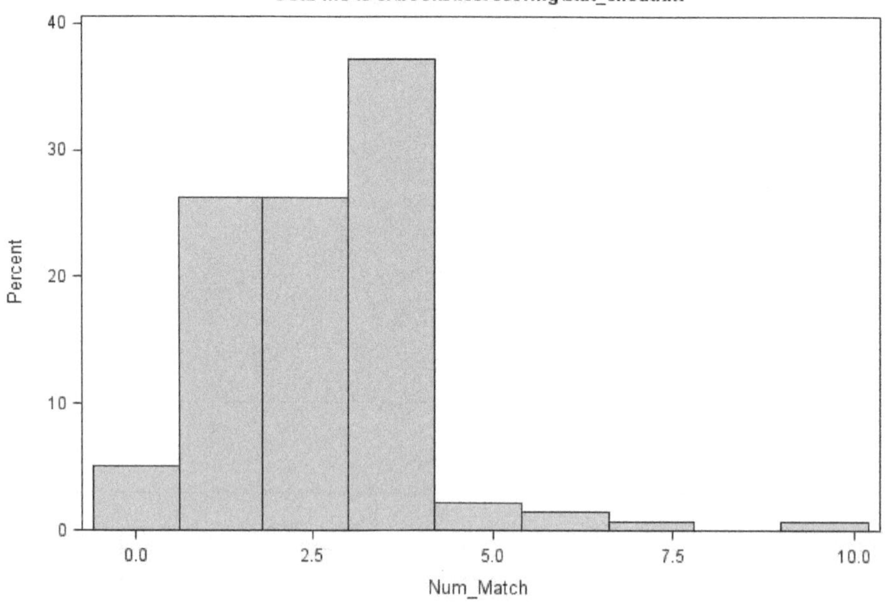


```
Exam file name: c:\books\test scoring\stat_cheat.txt
Number of Items: 56
Statistics for students 123456789 and 987654321


Student 123456789 got 11 items wrong
Students 123456789 and 987654321 have 10 wrong answers
in common
The mean number of matchs is 2.279
The standard deviation is 1.315
The z-score is  5.872 with a probability of 2.1533318E-
9
```

Results like this would lead you to the conclusion that there was cheating between students 123456789 and 987654321.

How to Detect Cheating: Method Two

An alternative method to detect cheating uses items that the two students in question both got wrong, known as joint-wrongs. The program that utilizes this method is similar to the previous one. You need to first find the set of joint-wrongs and then compute the number of the same wrong answers on this set of items for all students in the class. A program that uses this method is presented next (without detailed explanation):

Program 10.2: Using the Method of Joint-Wrongs to Detect Cheating

```
*Macro Joint_Wrong is similar to Compare_Wrong except that it uses
 joint-wrong items (items that both ID1 and ID2 got wrong) as the
basis
 for the calculations;

%macro Joint_Wrong
  (File=,          /*Name of text file containing key and
                     test data */
   Length_ID=,     /*Number of bytes in the ID          */
   Start=,         /*Starting column of student answers */
   ID1=,   /*ID of first student */
   ID2=,   /*ID of second student */
   Nitems= /*Number of items on the test */ );

    data ID_one(keep=ID Num_Wrong_One Ans_One1-Ans_One&Nitems
       Wrong_One1-Wrong_One&Nitems)
       ID_two(keep=ID Num_Wrong_Two Ans_Two1-Ans_Two&Nitems
       Wrong_Two1-Wrong_Two&Nitems)
       Others(keep=ID Num_wrong Ans1-Ans&Nitems Wrong1-Wrong&Nitems);
   /* Data set ID_One contains values for Student 1
      Data set ID_Two contains values for Student 2
      Data set Others contains data on other students
   */
      infile "&File" end=last pad;
   /*First record is the answer key*/
      array Ans[&Nitems] $ 1;
      array Ans_One[&Nitems] $ 1;
      array Ans_Two[&Nitems] $ 1;
      array Key[&Nitems] $ 1;
      array Wrong[&Nitems];
      array Wrong_One[&Nitems];
      array Wrong_Two[&Nitems];
      array Joint[&Nitems];
      retain Key1-Key&Nitems;
      if _n_ = 1 then input @&Start (Key1-Key&Nitems)($1.);
      input @1 ID $&Length_ID..
            @&Start (Ans1-Ans&Nitems)($1.);
```

```
      if ID = "&ID1" then do;
         do i = 1 to &Nitems;
            Wrong_One[i] = Key[i] ne Ans[i];
            Ans_One[i] = Ans[i];
         end;
         Num_Wrong_One = sum(of Wrong_One1-Wrong_One&Nitems);
         output ID_One others;
         return;
      end;
      if ID = "&ID2" then do;
         do i = 1 to &Nitems;
            Wrong_Two[i] = Key[i] ne Ans[i];
            Ans_Two[i] = Ans[i];
         end;
         Num_Wrong_Two = sum(of Wrong_Two1-Wrong_Two&Nitems);
         output ID_Two others;
         return;
      end;

      /*Compute wrong answers for the class, not including ID1 and ID2
*/
      Num_Wrong = 0;
      do i = 1 to &Nitems;
         Wrong[i] = Key[i] ne Ans[i];
      end;
      Num_Wrong = sum(of Wrong1-Wrong&Nitems);
      output Others;
   run;

*DATA step joint compute item number for the joint-wrongs;
   Data ID1ID2;
      array Wrong_One[&Nitems];
      array Wrong_Two[&Nitems];
      array Joint[&Nitems];
      set ID_One(keep=Wrong_One1-Wrong_One&Nitems);
      Set ID_Two(keep=Wrong_Two1-Wrong_Two&Nitems);
      Num_Wrong_Both = 0;
      do i = 1 to &Nitems;
         Joint[i] = Wrong_One[i] and Wrong_Two[i];
         Num_Wrong_Both + Wrong_One[i] and Wrong_Two[i];
      end;
      drop i;
   run;
```

```
*DATA step COMPARE counts the number of same wrong answers on joint-
wrongs.;
   data compare;
      if _n_ = 1 then do;
         set ID_One(keep=Ans_One1-Ans_One&Nitems);
         set ID1ID2;
      end;
      set others;
      array Ans[&Nitems] $ 1;
      array Ans_One[&Nitems] $ 1;
      array Joint[&Nitems];

      Num_Match = 0;
      do i = 1 to &Nitems;
         if Joint[i] = 1 then Num_Match + Ans[i] eq Ans_One[i];
      end;
      keep Num_Match ID;
   run;

   proc sgplot data=compare;
      title 'Distribution of the number of matches between';
      title2 "Students &ID1, &ID2, and the rest of the class";
      title3 "Data file is &File";
      histogram Num_Match;
   run;

   /*
   Compute the mean and standard deviation on the number of same
   wrong answers as ID1, but eliminate both ID1 and ID2 from the
   calculation
   */
   proc means data=compare(where=(ID not in ("&ID1" "&ID2")))noprint;
      var Num_Match;
      output out=Mean_SD mean=Mean_Num_match std=SD_Num_match;
   run;

     data _null_;
      file print;
      title1 "Exam file name: &File";
      title2 "Number of Items: &Nitems";
      title3 "Statistics for students &ID1 and &ID2";
      set mean_sd;
      set ID_One(Keep=Num_Wrong_One);
      set ID_Two(keep=Num_Wrong_Two);
      set ID1ID2(Keep=Num_Wrong_Both);
      set compare(where=(ID eq "&ID2"));
```

```
      Diff = Num_Wrong_Both - Mean_Num_Match;
      z = Diff / SD_Num_match;
      Prob = 1 - probnorm(z);

            put // "Student &ID1 has " Num_Wrong_One "items wrong" /
             "Student &ID2 has " Num_Wrong_Two "items wrong" /
             "Students &ID1 and &ID2 have " Num_Wrong_Both
             "wrong answers in common" /
             "Students &ID1 and &ID2 have " Num_Match
             "items with the same wrong answer" /
             73*'-' /
             "The mean number of matches is" Mean_Num_Match 6.3 /
             "The standard deviation is " SD_Num_match 6.3  /
             "The z-score is" z 6.3 " with a probability of " Prob;
   run;

   proc datasets library=work noprint;
      delete ID_One ID_Two Others compare ID1ID2 Mean_SD plot;
   quit;

%mend joint_wrong;
```

The program starts out by creating three data sets. Data sets ID_One and ID_Two contain one observation each with data for Students ID1 and ID2, respectively. Data set Others contains test data for the remaining students. Data set ID1ID2 computes the item numbers for the joint-wrongs. Data set Compare computes the number of the same wrong answers for the remaining students. PROC MEANS is used to compute the mean and standard deviation for the number of the same wrong answers (not including students ID1 and ID2). PROC SGPLOT is used to plot a histogram, and the final DATA _NULL_ step compute a *z*-score and *p*-value for the two students.

The following call routine computes statistics for the same test used in the previous example:

Calling the %Joint_Wrong Macro with IDs 123456789 and 987654321 Identified

```
%joint_wrong(File=c:\books\test scoring\stat_cheat.txt,
             Length_ID=9,
             Start=11,
             ID1=123456789,
             ID2=987654321,
             Nitems=56)
```

Output from %Joint_Wrong Macro

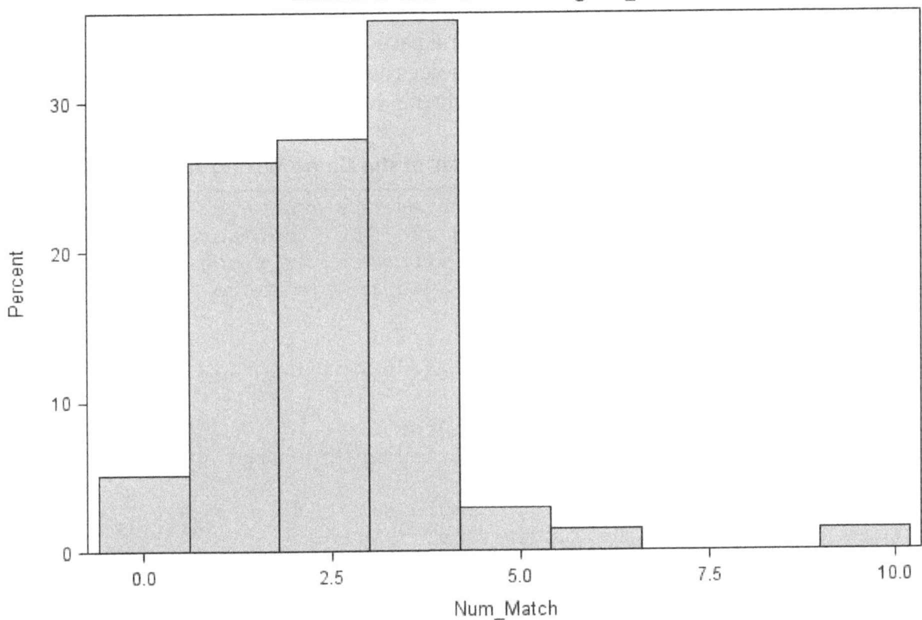


```
Exam file name: c:\books\test scoring\stat_cheat.txt
Number of Items: 56
Statistics for students 123456789 and 987654321


Student 123456789 has 11 items wrong
Student 987654321 has 12 items wrong
Students 123456789 and 987654321 have 10 wrong answers
in common
Students 123456789 and 987654321 have 10 items with the
same wrong answer
-------------------------------------------------------
------
The mean number of matches is 2.243
The standard deviation is  1.262
The z-score is 6.147 with a probability of 3.946199E-10
```

This method of cheating detection also produces convincing evidence that cheating occurred.

Searching for a Match

The last section in this chapter describes a program to search a file for wrong answer matches that suggest possible cheating. The program is very similar to Program 10.1, except that you supply a single ID and the program compares the wrong answer choices against all the other students in the class. Instead of computing a *z*-score and *p*-value for a particular student, this program lists all the students with a number of the same wrong answer choices as Student 1 where the probability is less than or equal to a predetermined threshold value. A listing of the program follows:

Program 10.3: Searching for IDs Where the Number of the Same Wrong Answers is Unlikely

```
*Macro Search searches a file and computes the number of same
 wrong answers as a student identified as ID1.  The macro outputs
 the z- and p-values for all students with similar wrong answer
 choices, with a p-value cutoff determined in the macro call;

%macro search
  (File=,         /*Name of text file containing key and
                    test data */
   Length_ID=,    /*Number of bytes in the ID          */
   Start=,        /*Starting column of student answers */
   ID1=,    /*ID of first student */
   Threshold=.01, /*Probability threshold */
   Nitems= /*Number of items on the test */);

***This DATA step finds the item numbers incorrect in the first ID;

    data ID_one(keep=ID Num_wrong_One Ans_One1-Ans_One&Nitems
       Wrong_One1-Wrong_One&Nitems)
       Not_One(keep=ID Num_wrong Ans1-Ans&Nitems Wrong1-Wrong&Nitems);
   /* Data set ID_One contains values for Student 1
      Data set Not_one contains data on other students
      Arrays with "one" in the variable names are data from
      ID1.
   */
   infile "&File" end=last pad;
   retain Key1-Key&&Nitems;
   /*First record is the answer key*/
      array Ans[&Nitems] $ 1;
      array Ans_One[&Nitems] $ 1;
      array Key[&Nitems] $ 1;
      array Wrong[&Nitems];
      array Wrong_One[&Nitems];
      if _n_ = 1 then input @&Start (Key1-Key&Nitems)($1.);
      input @1 ID $&Length_ID..
            @&Start (Ans1-Ans&Nitems)($1.);
```

```
      if ID = "&ID1" then do;
         do i = 1 to &Nitems;
            Wrong_One[i] = Key[i] ne Ans[i];
            Ans_One[i] = Ans[i];
         end;
         Num_Wrong_One = sum(of Wrong_One1-Wrong_One&Nitems);
         output ID_one;
         return;
      end;

      do i = 1 to &Nitems;
         Wrong[i] = Key[i] ne Ans[i];
      end;
      Num_Wrong = sum(of Wrong1-Wrong&Nitems);
      drop i;
      output Not_One;
   run;

   data compare;
      array Ans[&Nitems] $ 1;
      array Wrong[&Nitems];
      array Wrong_One[&Nitems];
      array Ans_One[&Nitems] $ 1;

      set Not_One;
      if _n_ = 1 then set ID_One(drop=ID);
     * if ID = "&ID" then delete;
      ***Compute # matches on set of wrong answers;
      Num_Match = 0;
      do i = 1 to &Nitems;
         if Wrong_One[i] = 1 then Num_Match + Ans[i] eq Ans_One[i];
      end;
      keep ID Num_Match Num_Wrong_One;
   run;

   proc means data=compare(where=(ID ne "&ID1")) noprint;
      var Num_Match;
      output out=means(drop=_type_ _freq_) mean=Mean_match
std=Sd_match;
   run;

   title 'Distribution of the number of matches between';
   title2 "Student &ID1 and the rest of the class";
   title3 "Data file is &File";
   proc sgplot data=compare;
      histogram Num_Match / binwidth=1;
   run;
```

```
   data _null_;
      file print;
      title "Statistics for student &ID1";
      if _n_ = 1 then set means;
      set compare;
      z = (Num_Match - Mean_match) / Sd_match;
      Prob = 1 - probnorm(z);

      if Prob < &Threshold  then
         put  /
         "ID = " ID "had " @20 Num_Match " wrong answers compare, "
                                "Prob = " Prob;

   run;

   proc datasets library=work noprint;
      delete compare ID_One means Not_One;
   quit;

%mend search;
```

The logic of this program is similar to Program 10.1, except that, in this case, you want to compute the *z*- and *p*-values for every student in the class and list those students where the *p*-value is below the threshold.

Let's first run this macro on the test_cheat.txt file. The macro call looks like this:

Calling the Macro to Detect Possible Cheaters in the test_cheat.txt File

```
%search(File=c:\books\test scoring\test_cheat.txt,
              Length_ID=3,
              Start=4,
              ID1=001,
              Threshold=.6,
              Nitems=6)
```

This calling sequence produces the following output:

Output from Program 10.8

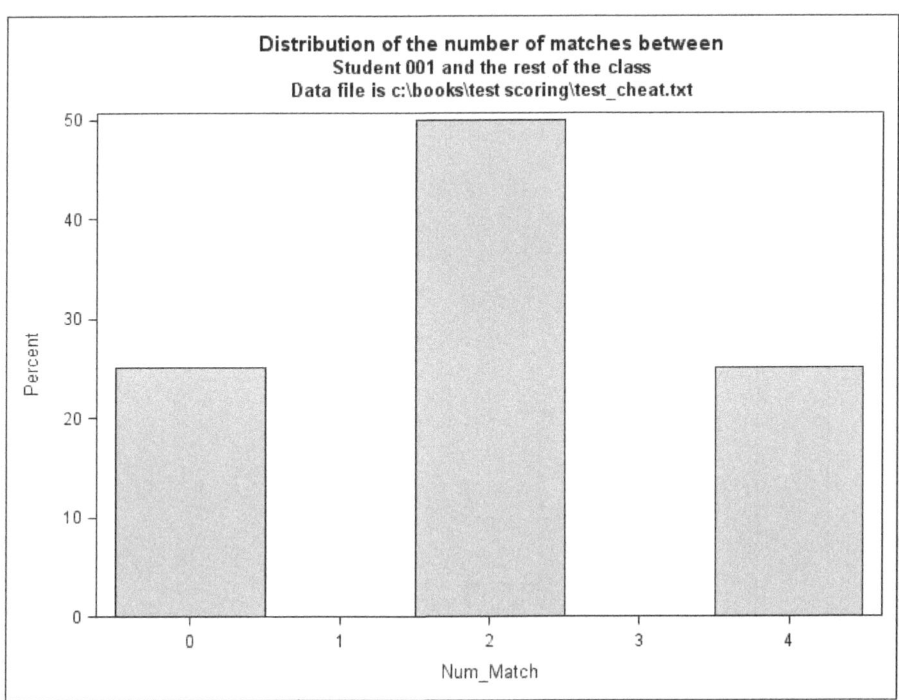


```
Statistics for student 001

ID = 003 had  2  wrong answers compare, Prob = 0.5

ID = 004 had  2  wrong answers compare, Prob = 0.5

ID = 005 had  4  wrong answers compare, Prob =
0.110335681
```

In normal practice, you would set the threshold value quite small. To see a more realistic example, let's run the macro on the stat_cheat.txt file like this:

Running the Search Macro on the stat_cheat.txt File

```
%search(File=c:\books\test scoring\stat_cheat.txt,
               Length_ID=9,
               Start=11,
               ID1=123456789,
               Threshold=.01,
               Nitems=56)
```

Here are the results:

Running the Search Macro on the stat_cheat.txt File with a Threshold of .01

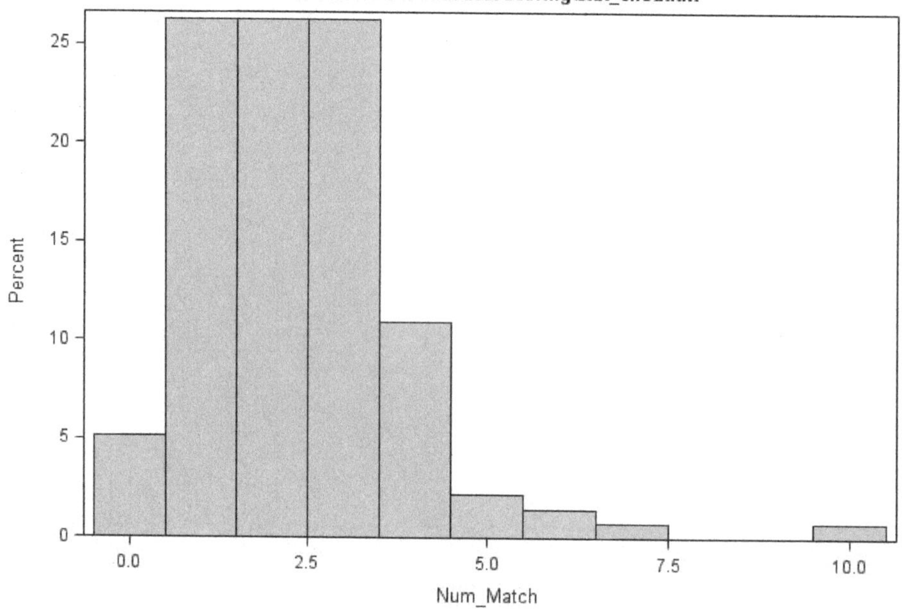


```
Statistics for student 123456789

ID = 987654321 had 10 wrong answers compare, Prob = 8.6804384E-8

ID = 957897193 had 7 wrong answers compare, Prob = 0.0007360208

ID = 605568642 had 6 wrong answers compare, Prob = 0.0062390769

ID = 700024487 had 6  wrong answers compare, Prob = 0.0062390769
```

The only highly significant match is with student 987654321. You may wonder why the probability for student 987654321 is larger than the probability computed by the first method described in this chapter. The reason is that in the first method, the mean and standard deviation for the number of the same wrong answers as Student 1 are computed with both Students 1 and 2 removed from the calculation. In the search program, Student 2 is included in the computation. Since Students 1 and 2 have so many of the same wrong answers, leaving Student 2 in the calculation inflates the mean and standard deviation. We suggest that if the search program identifies a possible occurrence of cheating, you should then run either the first program (Compare_Wrong) or the second program (Joint_Wrong) with the two students in question.

Conclusion

This chapter presented three programs (Compare_Wrong, Joint_Wrong, and Search) that you can use to investigate possible cheating on multiple-choice exams. It must be emphasized that great care needs to be taken when using these programs since the consequences are so important to the students in question and to the institution that is conducting the test.

References

Cody, Ron. 1985. "Statistical Analysis of Examinations to Detect Cheating." *Journal of Medical Education* Feb 60 (2) 136-137

Chapter 11: A Collection of Test Scoring, Item Analysis, and Related Programs

Introduction

This chapter contains programs (in the form of SAS macros) that perform most of the useful, test-related tasks described in this book. You can find explanations of how these programs work in the relevant chapters of this book. You will find an explanation of what each program does, how to run the program (including an example), and some sample output (if any).

You may obtain a copy of these programs from the following URL: support.sas.com/cody.

Scoring a Test (Reading Data from a Text File)

Program Name: Score_Text

Described in: Chapter 2

Purpose: To read raw test data from a text file, score the test, and create a SAS data set (temporary or permanent) to be used as input to other analysis programs. The output data set contains the student ID (ID), the answer key (Key1 to Key*n*), the student responses to each item (Ans1-Ans*n*), the scored variables (0=incorrect, 1=correct, variable names Score1-Score*n*), the raw score (Raw), and the percentage score (Percent).

Arguments:

File= the name of the text file containing the answer key and student answers.

Dsn= the name of the data set that contains the answer key, the student ID and student answers, and the scored items (0s and 1s). This can be a temporary (work) data set or a permanent SAS data set.

Length_ID= the number of characters (digits) in the student ID. The ID may contain any alphanumeric characters.

Start= the first column of the student ID.

Nitems= the number of items on the test.

Program 11.1: Score_Text

```
%macro Score_Text(File=, /*Name of the file containing the answer
                          key and the student answers       */
             Dsn=,       /* Name of SAS data set to create    */
             Length_ID=, /*Number of bytes in the ID          */
             Start=,     /*Starting column of student answers */
             Nitems=     /*Number of items on the test        */);

   data &Dsn;
      infile "&File" pad end=Last;
      array Ans[&Nitems] $ 1 Ans1-Ans&Nitems;     ***student Answers;
      array Key[&Nitems] $ 1 Key1-Key&Nitems;     ***Answer Key;
      array Score[&Nitems] 3 Score1-Score&Nitems; ***score array
                                                  1=right,0=wrong;
      retain Key1-Key&Nitems;
      if _n_ = 1 then input @&Start (Key1-Key&Nitems)($upcase1.);
      input @1  ID $&Length_ID..
            @&start (Ans1-Ans&Nitems)($upcase1.);
      do Item = 1 to &Nitems;
         Score[Item] = Key[Item] eq Ans[Item];
      end;
      Raw=sum (of Score1-Score&Nitems);
      Percent=100*Raw / &Nitems;
      keep Ans1-Ans&Nitems Key1-Key&Nitems Score1-Score&Nitems ID Raw
Percent;
      label ID      = 'Student ID'
            Raw     = 'Raw score'
            Percent = 'Percent score';
   run;

   proc sort data=&Dsn;
      by ID;
   run;
%mend score_text;
```

Sample call:

```
%score_text(file='c:\books\test scoring\stat_test.txt',
       dsn=score_stat,
       length_id=9,
       Start=11,
       Nitems=56)
```

Scoring a Test (Reading Data From an Excel File)

Program Name: Score_Excel

Described in: Chapter 2

Purpose: To score a test reading data from an Excel file.

Arguments:

Dsn= the name of the data set created by the Score_Excel macro.

Folder= the name of the folder where the Excel file is stored.

Worksheet= the name of the worksheet (Sheet1 is the default).

Nitems= the number of items on the test.

Program 11.2: Score_Excel

```
*Macro Score_Excel
 Purpose: To read test data from and Excel file and score the test;
/* This macro reads answer key and student answers stored in an
   Excel Worksheet.

   The first row of the worksheet contains labels as
   follows:
   Column A: ID
   Column B to Last column: Ans1 Ans2 etc.

   The second row of the worksheet contains the answer key
   starting in column B.
   The remaining rows contain the student ID in column A
   and the student answers in columns B to the end

 */
%macro Score_Excel(Dsn=, /* Name of SAS data set to create   */
   Folder=, /*Name of the folder where the worksheet is
             located                                         */
   File=,   /*Name of the Excel worksheet containing
            the answer key and the student answers           */
   Worksheet=Sheet1, /*worksheet name, Sheet1 is the default */
   Nitems=     /*Number of items on the test                 */);

   libname readxl "&folder\&File";

   data &Dsn;
      set readxl."&Worksheet$"n;
      retain Key1-Key&Nitems;
```

```
      array Key[&Nitems] $ 1;
      array Ans[&Nitems] $ 1;
      array Score[&Nitems] Score1-Score&Nitems;
      if _n_ = 1 then do Item = 1 to &Nitems;
         Key[Item] = Ans[Item];
      end;
      drop Item;
      do Item=1 to &Nitems;
         Score[Item] = Key[Item] eq Ans[Item];
      end;
      Raw=sum(of Score1-Score&Nitems);
      Percent=100*Raw / &Nitems;
      keep Ans1-Ans&Nitems Key1-Key&Nitems Score1-Score&Nitems ID Raw
Percent;
      label ID      = 'Student ID'
            Raw     = 'Raw Score'
            Percent = 'Percent Score';
   run;

   proc sort data=&Dsn;
      by ID;
   run;

%mend Score_Excel;
```

Sample Call

To help you understand exactly how you need to structure your worksheet, the Excel worksheet shown below was created for this example. It represents a five-item test with four students. Here it is:

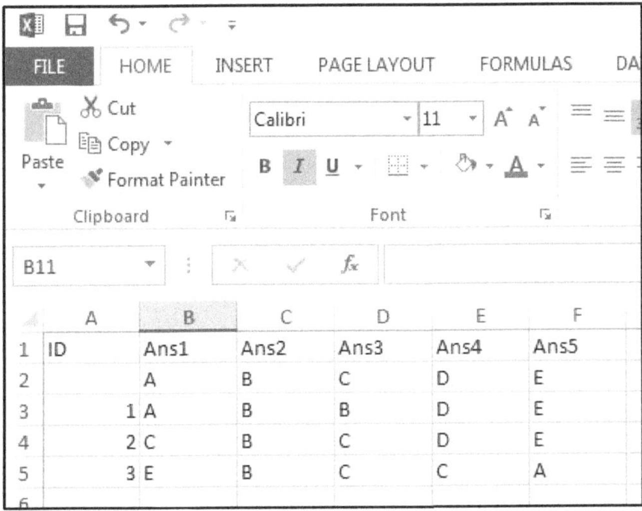

	A	B	C	D	E	F
1	ID	Ans1	Ans2	Ans3	Ans4	Ans5
2		A	B	C	D	E
3	1	A	B	B	D	E
4	2	C	B	C	D	E
5	3	E	B	C	C	A

Notice that the first row contains the labels ID and Ans1 to Ans5. The second row, starting in column 2, contains the answer key. The remaining rows contain the student IDs and student answers. The name of the spreadsheet was not changed from the default name Sheet1. Here is the call. (Note: If you are using a newer version of Excel that creates files with the XLSX extension, substitute that extension on the File= parameter.)

```
%score_excel(Folder=c:\books\test scoring,
             File=Sample Test.xls,
             Worksheet=Sheet1,
             Dsn=test,
             Nitems=5)
```

Printing a Roster

Program Name: Print_Roster

Described in: Chapter 2

Purpose: To print a student roster (Student ID and Score).

Arguments:

Dsn= the name of the data set created by the %Score_Text macro.

Program 11.3: Print_Roster

```
%macro print_roster(Dsn=  /*Data set name */);
   title "Roster of ID's, Raw and Percent Scores";
   title2 "Input data set is %upcase(&Dsn)";
   proc report data=&Dsn nowd headline;
      columns ID Raw Percent;
      define ID / "Student ID" width=10;
      define Raw / "Raw Score" width=5;
      define Percent / "Percent Score" width=7;
   run;
%mend print_roster;
```

Sample Call:

```
%print_roster(Dsn=Score_Stat)
```

Sample Output:

Roster of ID's, Raw and Percent Scores
Input data set is SCORE_STAT

Student ID	Raw Score	Percent Score
009917820	35	62.5
011413555	39	69.642857
012114262	40	71.428571
026052336	48	85.714286
033646387	41	73.214286
045841843	25	44.642857
051502147	41	73.214286
057010574	39	69.642857
057857387	39	69.642857
058904614	43	76.785714

Data Checking Program

Program Name: Data_Check

Described in: Chapter 4

Purpose: To identify data errors in student data (in a text file).

Arguments:

File= the name of the text file holding the answer key and student answers.

Length_ID= the number of characters (digits) in the student ID. The ID may contain any alphanumeric characters.

Start= the first column of the student ID.

Nitems= the number of items on the test.

Program 11.4: Data_Check

```
%macro data_check(File=, /*Name of the file containing the answer
                           key and the student answers        */
            Length_ID=, /*Number of bytes in the ID           */
            Start=,     /*Starting column of student answers */
            Nitems=     /*Number of items on the test         */);
   title "Checking for Invalid ID's and Incorrect Answer Choices";
   title2 "Input file is &File";
   data _null_;
      file print;
      if _n_ = 1 then put 54*'-' /;
      array ans[&Nitems] $ 1 Ans1-Ans&Nitems;
      infile 'c:\books\test scoring\data_errors.txt' pad;
      input @1  ID $&Length_ID..
            @11 (Ans1-Ans&Nitems)($upcase1.);
      if notdigit(ID) then put "Invalid ID " ID "in line " _n_;
      Do Item = 1 to &Nitems;
         if missing(Ans[Item]) then
         put "Item #" Item "Left Blank by Student " ID;
         else if Ans[Item] not in ('A','B','C','D','E') then
            put "Invalid Answer for Student " ID "for Item #"
            Item "entered as " Ans[Item];
      end;
   run;
%mend data_check;
```

Sample Call:

```
%data_check(File=c:\books\test scoring\data_errors.txt,
            Length_ID=9,
            Start=11,
            Nitems=10)
```

Sample Output:

Checking for Invalid ID's and Incorrect Answer Choices

Input file is c:\books\test scoring\data_errors.txt

```
--------------------------------------------------

Item #2 Left Blank by Student 123456789
Invalid Answer for Student 444444444 for Item #5 entered as ?
Invalid ID 777x77777 in line 9
```

Item Analysis Program

Program Name: Item

Described in: Chapter 5

Purpose: To perform item analysis, including answer frequencies, item difficulty, point-biserial correlation, and proportion correct by quartile. Note: You need to first score the test using either of the test scoring programs described earlier.

Arguments:

Dsn= the name of the data set created by the %Score_Text macro.

Nitems= the number of items on the test.

Program 11.5: Item

```
%macro Item(Dsn=,    /*Data set name */
           Nitems= /*Number of items on the test */);
   proc corr data=&Dsn nosimple noprint
            outp=corrout(where=(_type_='CORR'));
      var Score1-Score&Nitems;
      with Raw;
   run;
   ***reshape the data set;
   data corr;
      set corrout;
      array score[*] 3 Score1-Score&Nitems;
      Do i=1 to &Nitems;
         Corr = Score[i];
         output;
      end;
      keep i Corr;
   run;
   ***compute quartiles;
   proc rank data=&Dsn groups=4 out=quart(drop=Percent ID);
      ranks Quartile;
      var Raw;
   run;
   ***create item variable and reshape again;
   data tab;
      set quart;
      length Item $ 5 Quartile Correct i 3 Choice $ 1;
      array score[*] Score1-Score&Nitems;
      array ans{*} $ 1 Ans1-Ans&Nitems;
      array key{*} $ 1 Key1-Key&Nitems;
      Quartile = Quartile+1;
      Do i=1 to &Nitems;
         Item=right(put(i,3.)) || " " || Key[i];
         Correct=Score[i];
         Choice=Ans[i];
         output;
      end;
      keep i Item Quartile Correct Choice;
   run;
   proc sort data=tab;
      by i;
   run;
   ***combine correlations and quartile information;
   data both;
      merge corr tab;
      by i;
   run;
   ***print out a pretty table;
   options ls=132;
```

```
   title "Item Statistics";
   proc tabulate format=7.2 data=both order=internal noseps;
      label Quartile = 'Quartile'
            Choice   = 'Choices';
      class Item Quartile Choice;
      var Correct Corr;
      table Item='# Key'*f=6.,
      Choice*(pctn<Choice>)*f=3. Correct=' '*mean='Diff.'*f=Percent5.2
      Corr=' '*mean='Corr.'*f=5.2
      Correct=' '*Quartile*mean='Prop. Correct'*f=Percent7.2/
         rts=8;
      keylabel pctn='%' ;
   run;
   *Delete temporary data sets;
   proc datasets library=work noprint;
      delete corr;
      delete tab;
      delete both;
      delete corrout;
      delete quart;
   quit;
%mend Item;
```

Sample Call:

```
%Item(Dsn=Score_Stat, Nitems=56)
```

Sample Output:

Item Statistics

	Choices							Quartile			
	A	B	C	D	E			1	2	3	4
	%	%	%	%	%	Diff.	Corr.	Prop. Correct	Prop. Correct	Prop. Correct	Prop. Correct
# Key											
1 D	1	1	.	62	36	62%	0.49	22.6%	68.6%	67.5%	87.1%
2 C	1	5	82	1	10	82%	0.30	63.3%	71.4%	97.5%	93.5%
3 E	.	5	1	.	93	93%	0.32	80.6%	94.3%	97.5%	100%
4 B	17	46	18	13	7	46%	0.43	16.1%	35.3%	52.5%	80.6%
5 C	6	7	85	1	1	85%	0.35	64.5%	91.4%	90.0%	93.5%
6 B	1	96	3	.	.	96%	0.20	87.1%	97.1%	100%	96.8%
7 A	64	21	6	1	8	64%	0.42	32.3%	62.9%	65.0%	93.5%

Program to Delete Items and Rescore the Test

Program Name: Print_Roster

Described in: Chapter 6

Purpose: To delete items and rescore the test.

Arguments:

Dsn= the name of the data set created by the %Score_Text macro.

Dsn_Rescore= the name of the rescored data set.

Nitems= the number of items on the original test (before deleting items).

List= the list of item numbers to delete, separated by blanks.

Program 11.6: Print_Roster

```
*Macro Rescore
 Purpose: to rescore a test with a list of items to delete;
%macro Rescore(Dsn=, /*Name of data created by one of the scoring
macros */
               Dsn_Rescore=, /*Name of rescored data set */
               Nitems=, /*Number of itenms on the original test */
               List= /*List of items to delete, separated
                       by spaces */);
   *Note: One of the scoring macros must be run first;
   Data &Dsn_Rescore;
      set &Dsn(keep=ID Score1-Score&Nitems);
      array Score[&Nitems];
      retain Num &Nitems;
      Raw = 0;
      retain Num (&Nitems);
      *Determine the number of items after deletion;
      if _n_ = 1 then do i = 1 to &Nitems;
         if i in(&list) then Num = Num - 1;
      end;

      do i = 1 to &Nitems;
         Raw = Raw + Score[i]*(i not in (&list));
      end;
      drop i Num;
      Percent = 100*Raw/Num;
   run;
```

Sample Call:

```
%Rescore(Dsn=Score_stat, Dsn_Rescore=temp, Nitems=56, List=1 2 3 4 5)
```

Scoring Multiple Test Versions (Reading Test Data and Correspondence Data from Text Files)

Program Name: Mult_Versions_txt

Described in: Chapter 6

Purpose: To score multiple test versions.

Arguments:

Dsn= the name of the data set to create.

Length_ID= the number of characters (digits) in the student ID. The ID may contain any alphanumeric characters.

Start= the first column of the student ID.

Nversions= the number of versions of the test.

Nitems= the number of items on the test.

Corr_File= the name of the text file holding the correspondence information (see Chapter 6 for details).

Program 11.7: Mult_Versions_txt

```
*Macro Mult_Version_Txt
 Purpose: To read test and correspondence data for multiple
          test versions from text files and score the test;

%macro mult_version_txt(File=, /*Name of the file containing the
answer
                             key and the student answers        */
              Corr_File=, /*Name of the file with correspondence
                             information                        */
              Nversions=, /* Number of versions                 */
              Dsn=,       /* Name of SAS data set to create     */
              Length_ID=, /*Number of bytes in the ID           */
              Version_Col=, /* Column for version number        */
              Start=,     /*Starting column of student answers */
              Nitems=     /*Number of items on the test         */);
   data &Dsn;
      retain Key1-Key&Nitems;
      array Response[&Nitems] $ 1;
      array Ans[&Nitems] $ 1;
      array Key[&Nitems] $ 1;
      array Score[&Nitems];
      array Correspondence[&Nversions,&Nitems] _temporary_;
      if _n_ = 1 then do;
         *Load correspondence array;
         infile "&Corr_File";
         do Version = 1 to &Nversions;
            do Item = 1 to &Nitems;
               input Correspondence[Version,Item] @;
            end;
         end;
         infile "&File" pad;
         input @&Start (Key1-Key&Nitems) ($upcase1.);
      end;
      infile "&File" pad;
      input @1  ID $&Length_ID..
            @&Version_Col Version 1.
            @&start (Response1-Response&Nitems) ($Upcase1.);
      Raw = 0;
      do Item = 1 to &Nitems;
         Ans[Item] = Response[correspondence[Version,Item]];
         Score[Item] =  (Ans[Item] eq Key[Item]);
         Raw + Score[Item];
      end;
      drop Item Response1-Response&Nitems;
   run;

%mend mult_version_txt;
```

Sample Call:

```
%mult_version_txt(File=c:\books\test scoring\mult_versions.txt,
             Corr_File=c:\books\test scoring\corresp.txt,
             Nversions=3,
             Dsn=Multiple1,
             Length_ID=9,
             Version_Col=10,
             Start=11,
             Nitems=5)
```

Scoring Multiple Test Versions (Reading Test Data from a Text File and Correspondence Data from an Excel File)

Program Name: Mult_Versions_Excel_corr

Described in: Chapter 6

Purpose: To score multiple test versions.

Arguments:

File= the name of the text file containing the answer key and student data.

Corr_File= the name of the Excel file containing the correspondence data (see Chapter 6 for the structure of this file).

Nversions= the number of test versions.

Dsn= the name of the data set to create.

Length_ID= the number of characters (digits) in the student ID. The ID may contain any alphanumeric characters.

Start= the first column of the student ID.

Nitems= the number of items on the test.

Program 11.8: Mult_Versions_Excel_corr

```
*Macro Mult_Version_Excel_Corr
 Purpose: To read test and from a text file and correspondence
          data for multiple test versions from an Excel file
          and score the test;

%macro mult_version_Excel_corr
            (File=, /*Name of the file containing the answer
                             key and the student answers        */
             Corr_File=, /*Name of the Excel file with
                             correspondence information         */
             Nversions=, /* Number of versions                  */
             Dsn=,       /* Name of SAS data set to create      */
             Length_ID=, /*Number of bytes in the ID            */
             Version_Col=, /* Column for version number         */
             Start=,     /*Starting column of student answers */
             Nitems=     /*Number of items on the test         */);
   libname readxl "&Corr_File";
   data &Dsn;
      retain Key1-Key&Nitems;
      array Response[&Nitems] $ 1;
      array Ans[&Nitems] $ 1;
      array Key[&Nitems] $ 1;
      array Score[&Nitems];
      array Q[&Nitems];
      array Correspondence[&Nversions,&Nitems] _temporary_;
      if _n_ = 1 then do;
         *Load correspondence array;
         do Version = 1 to &Nversions;
           set readxl.'Sheet1$'n;
           do Item = 1 to &Nitems;
               Correspondence[Version,Item] = Q[Item];
            end;
         end;
         infile "&File" pad;
         input @&Start (Key1-Key&Nitems)($upcase1.);
      end;
      infile "&File" pad;
      input @1 ID $&Length_ID..
            @&Version_Col Version 1.
            @&Start (Response1-Response&Nitems)($upcase1.);
      Raw = 0;
```

```
      do Item = 1 to &Nitems;
         Ans[Item] = Response[correspondence[Version,Item]];
         Score[Item] =  (Ans[Item] eq Key[Item]);
         Raw + Score[Item];
      end;
      drop Item Response1-Response&Nitems;
   run;
%mend mult_version_Excel_corr;
```

Sample Call:

```
%mult_version_Excel_corr(File=c:\books\test scoring\mult_versions.txt,
             Corr_File=c:\books\test scoring\correspondence.xlsx,
             Nversions=3,
             Dsn=Multiple2,
             Length_ID=9,
             Version_Col=10,
             Start=11,
             Nitems=5)
```

Scoring Multiple Test Versions (Reading Test Data and Correspondence Data from Excel Files)

Program Name: Mult_Versions_Excel_corr

Described in: Chapter 6

Purpose: To score multiple test versions.

Arguments:

File= the name of the text file containing the answer key and student data.

Corr_File= the name of the Excel file containing the correspondence data (see Chapter 6 for the structure of this file).

Nversions= the number of test versions.

Dsn= the name of the data set to create.

Length_ID= the number of characters (digits) in the student ID. The ID may contain any alphanumeric characters.

Start= the first column of the student ID.

Nitems= the number of items on the test.

Program 11.9: Mult_Versions_Excel_corr

```
*Macro Mult_Version_Excel
 Purpose: To read test and from a text file and correspondence
          data for multiple test versions from an Excel file
          and score the test;

%macro mult_version_Excel
            (Test_File=, /*Name of the Excel file containing
                           the student answers                    */
             Corr_File=, /*Name of the Excel file with
                           correspondence information             */
             Nversions=, /* Number of versions                    */
             Dsn=,       /* Name of SAS data set to create        */
             Nitems=     /*Number of items on the test            */);

   libname readxl "&Corr_File";
   libname readtest "&Test_File";
   data &Dsn;
      retain Key1-Key&Nitems;
      array R[&Nitems] $ 1;
      array Ans[&Nitems] $ 1;
      array Key[&Nitems] $ 1;
      array Score[&Nitems];
      array Q[&Nitems];
      array Correspondence[&Nversions,&Nitems] _temporary_;
      if _n_ = 1 then do;
         *Load correspondence array;
         do Version = 1 to &Nversions;
           set readxl.'Sheet1$'n;
           do Item = 1 to &Nitems;
               Correspondence[Version,Item] = Q[Item];
            end;
         end;
         set readtest.'Sheet1$'n(rename=(ID=Num_ID));
         do Item = 1 to &Nitems;
            Key[Item] = R[Item];
         end;
      end;
```

```
      set readtest.'Sheet1$'n (firstobs=2);
      Raw = 0;
      do Item = 1 to &Nitems;
         Ans[Item] = R[correspondence[Version,Item]];
         Score[Item] =  (Ans[Item] eq Key[Item]);
         Raw + Score[Item];
      end;
      drop Item R1-R&Nitems Num_ID Q1-Q&Nitems;
   run;

%mend mult_version_Excel;
```

Sample Call:

```
%mult_version_Excel
            (Test_File=c:\books\test scoring\Mult_Versions.xlsx,
             Corr_File=c:\books\test scoring\correspondence.xlsx,
             Nversions=3,
             Dsn=Multiple3,
             Nitems=5)
```

KR-20 Calculation

Program Name: KR_20

Described in: Chapter 7

Purpose: To perform item analysis, including answer frequencies, item difficulty, point-biserial correlation, and proportion correct by quartile.

Arguments:

Dsn= the name of the data set created by the %Score_Text macro.

Nitems= the number of items on the test.

Program 11.10: KR_20

```
%macro KR_20(Dsn=,    /*Name of the data set */
             Nitems= /*Number of items on the test */);
   *Note: You must run one of the the Score macros before running this
macro;
   proc means data=&dsn noprint;
      output out=variance var=Raw_Var Item_Var1-Item_Var&Nitems;
      var Raw Score1-Score&Nitems;
   run;

   title "Computing the Kuder-Richard Formula 20";
   data _null_;
      file print;
      set variance;
      array Item_Var[&Nitems];
      do i = 1 to &Nitems;
         Item_Variance + Item_Var[i];
      end;
      KR_20 = (&Nitems/%eval(&Nitems - 1))*(1 -
Item_Variance/Raw_Var);
      put (KR_20 Item_Variance Raw_Var)(= 7.3);
      drop i;
   run;

   proc datasets library=work noprint;
      delete variance;
   quit;
%mend KR_20;
```

Sample Call:

```
%KR_20(Dsn=Score_Stat, Nitems=56)
```

Sample Output:

Computing the Kuder-Richard Formula 20

```
KR_20=0.798 Item_Variance=7.902 Raw_Var=36.551
```

Program to Detect Cheating (Method One)

Program Name: Compare_Wrong

Described in: Chapter 10

Purpose: To detect cheating on a multiple-choice exam. This method utilizes the set of wrong answers from one student and counts the number of the same wrong answers in this set to all other students in the class.

Arguments:

File= the name of the text file holding the answer key and student answers.

Length_ID= the number of characters (digits) in the student ID. The ID may contain any alphanumeric characters.

Start= the first column of the student ID.

ID1= the ID of the first student.

ID2= the ID of the second student.

Nitems= the number of items on the test.

Program 11.11: Compare_Wrong

```
%macro Compare_wrong
  (File=,          /*Name of text file containing key and
                     test data */
   Length_ID=,     /*Number of bytes in the ID          */
   Start=,         /*Starting column of student answers */
   ID1=,   /*ID of first student */
   ID2=,   /*ID of second student */
   Nitems= /*Number of items on the test */ );

    data ID_one(keep=ID Num_wrong_One Ans_One1-Ans_One&Nitems
Wrong_One1-Wrong_One&Nitems)
       Not_One(keep=ID Num_wrong Ans1-Ans&Nitems Wrong1-Wrong&Nitems);
   /* Data set ID_One contains values for Student 1
      Data set Not_one contains data on other students
      Arrays with "one" in the variable names are data from
      ID1.
   */
      infile "&File" end=last pad;
   /*First record is the answer key*/
      array Ans[&Nitems] $ 1;
      array Ans_One[&Nitems] $ 1;
      array Key[&Nitems] $ 1;
```

```
      array Wrong[&Nitems];
      array Wrong_One[&Nitems];
      retain Key1-Key&Nitems;
      if _n_ = 1 then input @&Start (Key1-Key&Nitems)($upcase1.);
      input @1 ID $&Length_ID..
            @&Start (Ans1-Ans&Nitems)($upcase1.);
      if ID = "&ID1" then do;
         do i = 1 to &Nitems;
            Wrong_One[i] = Key[i] ne Ans[i];
            Ans_One[i] = Ans[i];
         end;
         Num_Wrong_One = sum(of Wrong_One1-Wrong_One&Nitems);
         output ID_one;
         return;
      end;

      do i = 1 to &Nitems;
         Wrong[i] = Key[i] ne Ans[i];
      end;
      output Not_One;
   run;

   /*
   DATA step COMPARE counts the number of same wrong answers as
student
   ID1.
   */
   data compare;
      if _n_ = 1 then set ID_One;
      set Not_One;
      array Ans[&Nitems] $ 1;
      array Ans_One[&Nitems] $ 1;
      array Wrong[&Nitems];
      array Wrong_One[&Nitems];
      Num_Match = 0;
      do i = 1 to &Nitems;
         if Wrong_One[i] = 1 then Num_Match + Ans[i] eq Ans_One[i];
      end;
      keep Num_Match ID;
   run;
```

```
   proc sgplot data=compare;
      title 'Distribution of the number of matches between';
      title2 "Students &ID1, &ID2, and the rest of the class";
      title3 "Data file is &File";
      histogram Num_Match;
   run;

   /*
   Compute the mean and standard deviation on the number of the same
   wrong answers as ID1 but eliminate both ID1 and ID2 from the
   calculation
   */
   proc means data=compare(where=(ID not in ("&ID1" "&ID2")))noprint;
      var Num_Match;
      output out=Mean_SD mean=Mean_Num_match std=SD_Num_match;
   run;

   data _null_;
      file print;
      title1 "Exam file name: &File";
      title2 "Number of Items: &Nitems";
      title3 "Statistics for students &ID1 and &ID2";
      set compare (where=(ID = "&ID2"));
      set mean_sd;
      set ID_One;

      Diff = Num_Match - Mean_Num_Match;
      z = Diff / SD_Num_match;
      Prob = 1 - probnorm(z);

      put // "Student &ID1 got " Num_wrong_One "items wrong" /
             "Students &ID1 and &ID2 have " Num_Match "wrong answers
in common" /
             "The mean number of matches is" Mean_Num_Match 6.3/
             "The standard deviation is" SD_Num_match  6.3/
             "The z-score is " z 6.3 " with a probability of " Prob;
   run;

   proc datasets library=work noprint;
      delete ID_One;
      delete Not_One Means;
   quit;

%mend compare_wrong;
```

Sample Call:

```
%compare_wrong(File=c:\books\test scoring\stat_cheat.txt,
               Length_ID=9,
               Start=11,
               ID1=123456789,
               ID2=987654321,
               Nitems=56)
```

Sample Output:

Exam file name: c:\books\test scoring\stat_cheat.txt
Number of Items: 56
Statistics for students 123456789 and 987654321

```
Student 123456789 got 11 items wrong
Students 123456789 and 987654321 have 10 wrong answers in common
The mean number of matches is 2.279
The standard deviation is 1.315
The z-score is  5.872 with a probability of 2.1533318E-9
```

Program to Detect Cheating (Method Two)

Program Name: Joint_Wrong

Described in: Chapter 10

Purpose: To detect cheating on a multiple-choice exam. This method utilizes the set of wrong answers that two students both got wrong (called *joint-wrongs*) and counts the number of the same wrong answers in this set to all other students in the class.

Arguments:

File= the name of the text file holding the answer key and student answers.

Length_ID= the number of characters (digits) in the student ID. The ID may contain any alphanumeric characters.

Start= the first column of the student ID.

ID1= the ID of the first student.

ID2= the ID of the second student.

Nitems= the number of items on the test.

Program 11.12: Joint_Wrong

```
%macro Joint_Wrong
  (File=,          /*Name of text file containing key and
                     test data */
   Length_ID=,     /*Number of bytes in the ID          */
   Start=,         /*Starting column of student answers */
   ID1=,    /*ID of first student */
   ID2=,    /*ID of second student */
   Nitems= /*Number of items on the test */ );

    data ID_one(keep=ID Num_Wrong_One Ans_One1-Ans_One&Nitems
Wrong_One1-Wrong_One&Nitems)
         ID_two(keep=ID Num_Wrong_Two Ans_Two1-Ans_Two&Nitems
Wrong_Two1-Wrong_Two&Nitems)
         Others(keep=ID Num_wrong Ans1-Ans&Nitems Wrong1-
Wrong&Nitems);
   /* Data set ID_One contains values for Student 1
      Data set ID_Two contains values for Student 2
      Data set Others contains data on other students
   */
      infile "&File" end=last pad;
   /*First record is the answer key*/
      array Ans[&Nitems] $ 1;
      array Ans_One[&Nitems] $ 1;
      array Ans_Two[&Nitems] $ 1;
      array Key[&Nitems] $ 1;
      array Wrong[&Nitems];
      array Wrong_One[&Nitems];
      array Wrong_Two[&Nitems];
      array Joint[&Nitems];
      retain Key1-Key&Nitems;
      if _n_ = 1 then input @&Start (Key1-Key&Nitems)($upcase1.);
      input @1 ID $&Length_ID..
            @&Start (Ans1-Ans&Nitems)($upcase1.);
      if ID = "&ID1" then do;
         do i = 1 to &Nitems;
            Wrong_One[i] = Key[i] ne Ans[i];
            Ans_One[i] = Ans[i];
         end;
         Num_Wrong_One = sum(of Wrong_One1-Wrong_One&Nitems);
         output ID_One others;
         return;
      end;
      if ID = "&ID2" then do;
         do i = 1 to &Nitems;
            Wrong_Two[i] = Key[i] ne Ans[i];
            Ans_Two[i] = Ans[i];
         end;
```

```
         Num_Wrong_Two = sum(of Wrong_Two1-Wrong_Two&Nitems);
         output ID_Two others;
         return;
      end;

/*Compute wrong answers for the class not including ID1 and ID2 */
      Num_Wrong = 0;
      do i = 1 to &Nitems;
         Wrong[i] = Key[i] ne Ans[i];
      end;
      Num_Wrong = sum(of Wrong1-Wrong&Nitems);
      output Others;
   run;

*DATA step joint compute item number for the joint-wrongs;
   Data ID1ID2;
      array Wrong_One[&Nitems];
      array Wrong_Two[&Nitems];
      array Joint[&Nitems];
      set ID_One(keep=Wrong_One1-Wrong_One&Nitems);
      Set ID_Two(keep=Wrong_Two1-Wrong_Two&Nitems);
      Num_Wrong_Both = 0;
      do i = 1 to &Nitems;
         Joint[i] = Wrong_One[i] and Wrong_Two[i];
         Num_Wrong_Both + Wrong_One[i] and Wrong_Two[i];
      end;
      drop i;
   run;

*DATA step COMPARE counts the number of same wrong answers on joint-
wrongs.;
   data compare;
      if _n_ = 1 then do;
         set ID_One(keep=Ans_One1-Ans_One&Nitems);
         set ID1ID2;
      end;
      set others;
      array Ans[&Nitems] $ 1;
      array Ans_One[&Nitems] $ 1;
      array Joint[&Nitems];

      Num_Match = 0;
      do i = 1 to &Nitems;
         if Joint[i] = 1 then Num_Match + Ans[i] eq Ans_One[i];
      end;
      keep Num_Match ID;
   run;
```

```
   proc sgplot data=compare;
      title 'Distribution of the number of matches between';
      title2 "Students &ID1, &ID2, and the rest of the class";
      title3 "Data file is &File";
      histogram Num_Match;
   run;

   /*
   Compute the mean and standard deviation on the number of same
   wrong answers as ID1 but eliminate both ID1 and ID2 from the
   calculation
   */
   proc means data=compare(where=(ID not in ("&ID1" "&ID2")))noprint;
      var Num_Match;
      output out=Mean_SD mean=Mean_Num_match std=SD_Num_match;
   run;

     options ls=132;
     data _null_;
      file print;
      title1 "Exam file name: &File";
      title2 "Number of Items: &Nitems";
      title3 "Statistics for students &ID1 and &ID2";
      set mean_sd;
      set ID_One(Keep=Num_Wrong_One);
      set ID_Two(keep=Num_Wrong_Two);
      set ID1ID2(Keep=Num_Wrong_Both);
      set compare(where=(ID eq "&ID2"));

      Diff = Num_Wrong_Both - Mean_Num_Match;
      z = Diff / SD_Num_match;
      Prob = 1 - probnorm(z);

      put // "Student &ID1 has " Num_Wrong_One "items wrong" /
             "Student &ID2 has " Num_Wrong_Two "items wrong" /
             "Students &ID1 and &ID2 have " Num_Wrong_Both "wrong
answers in common" /
             "Students &ID1 and &ID2 have " Num_Match "items with the
same wrong answer" /
             73*'-' /
             "The mean number of matches is" Mean_Num_Match 6.3 /
             "The standard deviation is " SD_Num_match 6.3  /
             "The z-score is" z 6.3 " with a probability of " Prob;
   run;
```

```
   proc datasets library=work noprint;
      delete ID_One ID_Two Others compare ID1ID2 Mean_SD plot;
   quit;

%mend joint_wrong;
```

Sample Call:

```
%joint_wrong(File=c:\books\test scoring\stat_cheat.txt,
             Length_ID=9,
             Start=11,
             ID1=123456789,
             ID2=987654321,
             Nitems=56)
```

Sample Output:

Exam file name: c:\books\test scoring\stat_cheat.txt
Number of Items: 56
Statistics for students 123456789 and 987654321

```
Student 123456789 has 11 items wrong
Student 987654321 has 12 items wrong
Students 123456789 and 987654321 have 10 wrong answers in common
Students 123456789 and 987654321 have 10 items with the same wrong answer
-----------------------------------------------------------------------
The mean number of matches is 2.243
The standard deviation is  1.262
The z-score is 6.147 with a probability of 3.946199E-10
```

Program to Search for Possible Cheating

Program Name: Search

Described in: Chapter 10

Purpose: To detect cheating on a multiple-choice exam. This method utilizes the set of wrong answers that two students both got wrong (called joint-wrongs) and counts the number of the same wrong answers in this set to all other students in the class.

Arguments:

File= the name of the text file holding the answer key and student answers.

Length_ID= the number of characters (digits) in the student ID. The ID may contain any alphanumeric characters.

Start= the first column of the student ID.

ID1= the ID of the first student.

Threshold= the *p*-value cutoff. All students with the same number of wrong answers as Student 1 that result in a *p*-value below the cutoff will be listed.

Nitems= the number of items on the test.

Program 11.13: Search

```
%macro search
  (File=,           /*Name of text file containing key and
                      test data */
   Length_ID=,      /*Number of bytes in the ID          */
   Start=,          /*Starting column of student answers */
   ID1=,   /*ID of first student */
   Threshold=.01, /*Probability threshold */
   Nitems= /*Number of items on the test */ );

***This Data Step finds the item numbers incorrect in the first id;

    data ID_one(keep=ID Num_wrong_One Ans_One1-Ans_One&Nitems
Wrong_One1-Wrong_One&Nitems)
       Not_One(keep=ID Num_wrong Ans1-Ans&Nitems Wrong1-Wrong&Nitems);
   /* Data set ID_One contains values for Student 1
      Data set Not_one contains data on other students
      Arrays with "one" in the variable names are data from
      ID1.
   */
   infile "&File" end=last pad;
   retain Key1-Key&&Nitems;
   /*First record is the answer key*/
      array Ans[&Nitems] $ 1;
      array Ans_One[&Nitems] $ 1;
      array Key[&Nitems] $ 1;
      array Wrong[&Nitems];
      array Wrong_One[&Nitems];
      if _n_ = 1 then input @&Start (Key1-Key&Nitems)($upcase1.);
      input @1 ID $&Length_ID..
            @&Start (Ans1-Ans&Nitems)($upcase1.);
      if ID = "&ID1" then do;
         do i = 1 to &Nitems;
            Wrong_One[i] = Key[i] ne Ans[i];
            Ans_One[i] = Ans[i];
         end;
```

```
         Num_Wrong_One = sum(of Wrong_One1-Wrong_One&Nitems);
         output ID_one;
         return;
      end;

      do i = 1 to &Nitems;
         Wrong[i] = Key[i] ne Ans[i];
      end;
      Num_Wrong = sum(of Wrong1-Wrong&Nitems);
      drop i;
      output Not_One;
   run;

   data compare;
      array Ans[&Nitems] $ 1;
      array Wrong[&Nitems];
      array Wrong_One[&Nitems];
      array Ans_One[&Nitems] $ 1;

      set Not_One;
      if _n_ = 1 then set ID_One(drop=ID);
     * if ID = "&ID" then delete;
      ***Compute # matches on set of wrong answers;
      Num_Match = 0;
      do i = 1 to &Nitems;
         if Wrong_One[i] = 1 then Num_Match + Ans[i] eq Ans_One[i];
      end;
      keep ID Num_Match Num_Wrong_One;
   run;

   proc means data=compare(where=(ID ne "&ID1")) noprint;
      var Num_Match;
      output out=means(drop=_type_ _freq_) mean=Mean_match
std=Sd_match;
   run;

   title 'Distribution of the number of matches between';
   title2 "Student &ID1 and the rest of the class";
   title3 "Data file is &File";

   proc sgplot data=compare;
      histogram Num_Match / binwidth=1;
   run;
```

```
   data _null_;
      file print;
      title "Statistics for student &ID1";
      if _n_ = 1 then set means;
      set compare;
      z = (Num_Match - Mean_match) / Sd_match;
      Prob = 1 - probnorm(z);

      if Prob < &Threshold  then
          put  /
          "ID = " ID "had " Num_Match " wrong answers compare, "
                              "Prob = " Prob;

   run;

   proc datasets library=work noprint;
      delete compare ID_One means Not_One;
   quit;

%mend search;
```

Sample Call:

```
%search(File=c:\books\test scoring\stat_cheat.txt,
               Length_ID=9,
               Start=11,
               ID1=123456789,
               Threshold=.01,
               Nitems=56)
```

Sample Output:

Statistics for student 123456789

```
ID = 987654321 had 10  wrong answers compare, Prob = 8.6804384E-8

ID = 957897193 had 7  wrong answers compare, Prob = 0.0007360208

ID = 605568642 had 6  wrong answers compare, Prob = 0.0062390769

ID = 700024487 had 6  wrong answers compare, Prob = 0.0062390769
```

Conclusion

The programs (macros) in this chapter should, hopefully, allow you to perform a wide variety of test scoring and item analysis activities. You may use them as is or, if you have some SAS programming skills, modify them for your own use.

Index

F

G

H

I

J

K

L

M

N

O

P

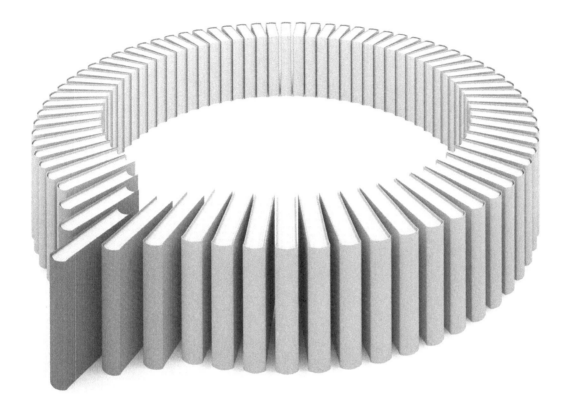

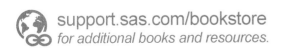

CPSIA information can be obtained at www.ICGtesting.com
Printed in the USA
LVOW01s2234180315

431135LV00005B/31/P